T0296087

FOREST, STEPPE
& TUNDRA

FOREST, STEPPE & TUNDRA

STUDIES IN ANIMAL ENVIRONMENT

by

MAUD D. HAVILAND

(MRS H. H. BRINDLEY)

SOMETIME FELLOW OF NEWNHAM COLLEGE, AND
HONORARY MEMBER (WOMAN) OF THE
BRITISH ORNITHOLOGISTS' UNION

CAMBRIDGE

AT THE UNIVERSITY PRESS

MCMXXVI

CAMBRIDGE
UNIVERSITY PRESS

University Printing House, Cambridge CB2 8BS, United Kingdom

Cambridge University Press is part of the University of Cambridge.

It furthers the University's mission by disseminating knowledge in the pursuit of education, learning and research at the highest international levels of excellence.

www.cambridge.org
Information on this title: www.cambridge.org/9781107455573

© Cambridge University Press 1926

First published 1926
First paperback edition 2014

A catalogue record for this publication is available from the British Library

ISBN 978-1-107-45557-3 Paperback

TO
THE MEMORY OF
DORA CURTIS
AND THE DAYS AND NIGHTS
WE SPENT TOGETHER
IN FOREST, STEPPE
AND TUNDRA

PREFACE

THESE essays are based upon a course of lectures given at Cambridge in 1924. In their preparation I have drawn freely upon the work of others to knit together my own scattered observations and impressions; and where the source of such assistance is not already indicated in the text, I should like to make full acknowledgment here. I am much indebted to Mrs Elizabeth Hewitt for permission to reproduce the curves on pages 25 and 27 from *The Conservation of the Wild Life of Canada*, by the late Dr C. Gordon Hewitt; and to Mr W. G. Kendrew for permission to extract statistics from his book, *The Climates of the Continents*, for the temperature and rainfall curves on pages 38, 120, and 198. I am obliged to Messrs George Phillips for the use of their outline map of Asia, but for the details of distribution, etc., in the latter, I alone am responsible. The quotation on p. 156 is by courtesy of Messrs Mills and Boon.

It is a pleasant duty to offer my thanks to Mr F. A. Potts for much helpful criticism, and for his advice in the arrangement of parts of the text. I must also thank Mr T. A. Coward, who was good enough to read and discuss the original manuscript with me, and my husband for his help in reading the proofs. Lastly, I should like to acknowledge my debt to Mrs Chadwick, whose enthusiasm for her own studies and sympathetic interest in those of her friends have been a constant stimulus to the completion of this book.

MAUD D. BRINDLEY

CONTENTS

LIST OF ILLUSTRATIONS

INTRODUCTION

LIFE has been defined as "a continual struggle with the environment"; and the cynic who, from the human point of view, summed it up as "one damned thing after another" was only making the statement in another way. Throughout existence, every organism is swayed by two contending forces—one innate and conservative, urging it to appear and act as its ancestors appeared and acted: the other environmental and progressive, compelling it to change in conformity with its surroundings. The animal we know is the product of an age-long struggle to reconcile constitutional limitations with environmental exigencies.

For the œcologist, the environment includes every external influence which surrounds the living being—the climate, the soil, the topography of the country and its flora and fauna. It is the setting in which the animal or plant is placed, "the mould into which the organism fits" (38). We owe to Haeckel the useful word "Œcology" for the study of the relations of this environmental mould and its occupants; and nowadays the title "Bionomics" is sometimes given to the reactions of habit and life history to the same causes[1].

It is for the zoologist in the practice of his science to determine the part played in the moulding of the organism by the two opposing forces of heredity and environment; and he is confronted by this problem as soon as he seeks to devise a natural classification of animals. The existing systems are taxonomic, that is to say they are based on phylogenetic considerations, and express our views of the evolution of species and their morphological affinities. This is the sound foundation for a system of classification, because, while all biologists share the belief that existing forms have arisen from a common stock, the part played by the environment in this evolution is still in dispute. It is often difficult to distinguish between characters of

---

[1] The good old term "Natural History" has fallen into disuse, but might well be revived to include much of what is now called Bionomics.

taxonomic value and those which are environmental adaptations. Sometimes allied forms differ considerably in adaptation, and in some cases adaptational characters may assume taxonomic importance. This difficulty has so impressed some modern œcologists that, for the purposes of their study, they have proposed schemes of classification based on bionomical or physiological considerations. Such a system has its convenience in the detailed œcological survey of limited areas, and in the use of special terms to express associations of organisms possessing similar reactions to a common environmental complex, but its employment in a more general or exact way is not practicable in the present work. The taxonomist considers life from the angle of what it is and was: the œcologist wishes to study what it does and did. Data for the first point of view are available, and can be tested by more or less precise methods. Data for the second are not available, at any rate at present, and even if we possessed them, it is doubtful whether they would readily lend themselves to comparison and measurement. It is true that, within limits, reactions to environmental stimuli can be distinguished and recorded. It is possible for instance to test the reactions of certain organisms to CO_2, or to increased insolation or humidity; but a general classification based on such a primary reaction is out of the question, because it would be impossible to subject the whole animal kingdom to experiment. Such experiments are of high importance and interest, but taxonomy remains the sheet-anchor for the explorer who ventures into the œcological sea. If an œcological "classification" is necessary at all, it is preferable to group the different animals of a region by a common link, such as food, into a series of bionomical complexes as will be described later on. This would substitute for the phylogenetic tree that familiar figure of taxonomy, the concept of a net or woven cloth of intricate pattern, whose threads, no matter what their colour or size, are all mutually interdependent and essential to the strength and design of the whole fabric.

Botanists however sometimes use a form of œcological classification when they approach their subject from the point of view of environment. Thus Schimper, Warming and others write of guilds, formations, associations and so on; and Warming (45) in this sense regards the *formation* and the *association* as the œcological equivalents of the taxonomic genus and species respectively. Such conceptions and terms can be used with greater precision by the botanist than by the zoologist because of the essential difference between plants and animals. The plant is fixed, and can only meet change in the surrounding conditions by structural and physiological changes in itself. The animal is mobile, and within limits can change its environment to suit its needs by moving elsewhere.

Thus an apple tree and a swallow must both prepare for the winter. The tree sheds its leaves and lowers its metabolism. The bird migrates to Africa. In a sense the plant is the captive of its environment, and wears its chains—the structural adaptations to its surroundings—plainly for all to see. On the other hand, the animal may be compared to a prisoner on bail or ticket-of-leave. Its movements are restricted, but the fetters are less obvious. In fact the physical surroundings can be best deduced from what the plant *is*, and from what the animal *does*. Shelford (37) remarks that for œcological purposes, it is the behaviour (bionomics) of animals, and not, as is often thought, their structure which should be compared with structure in the vegetable kingdom; and he also points out that again it is behaviour which offers the truest analogy to the development of human culture.

It is the faithful correlation of plant growth with the physical environment, especially to the important factor, or complex of factors, called "climate," that leads us naturally to define the main types of land environment in terms of plant life as Woodland, Grassland and Desert. These three great divisions have meaning not only for the botanist, but for others also; and this is easily understood, for vegetation is the apparel of scenery. As Darwin wrote: "A traveller should be a botanist, for in all views plants form the chief embellishment."

But when the zoologist, forsaking botanical terms, tries to classify environments in the language of his own science, he cannot construct a workable scheme. Here and there a region, or zone of a region, is characterised by a special type of animal life; but sooner or later the classification breaks down, and he finds that he must fall back on the language of the botanist or geologist. Nevertheless, though these great formations cannot be expressed in terms of animal life, they represent real though often ill-defined conceptions to the zoologist. In certain kinds of country—forests, steppes, mountains, deserts—certain types of animals predominate, while others are rare or absent; and it is often found that the predominant forms possess a certain number of adaptations in common.

This was brought home to me more than once between the years 1914 and 1922 when it was my good fortune to visit extreme examples of Woodland, Grassland and Desert in different parts of the world; and this book is the outcome of the lack that I myself then felt of some guide to the general trend of animal life amid surroundings that were new to me and presented unexpected œcological complications and problems. For although the large formations, which are sometimes called climatic or geographic, are typical of the country over wide areas, they are not homogeneous. Each is interspersed with smaller local or edaphic formations, due to accidents of soil or topography. We speak of the Sahara or Gobi deserts as sandy wastes, although we are well aware that the typical desert scenery is interrupted by oases and watercourses. Rivers, swamps, lakes, mountains, forests, clearings, etc., by modifying the soil, rainfall, and other environmental factors, all give rise to local formations which are of great importance to the distribution and development of the flora and fauna. In fact the œcological study of any region is essentially the study of local *versus* climatic conditions.

For œcological purposes, woodland includes all the forested regions of the earth, from the stunted birch and conifer tracts of the north, or the deciduous woods of the temperate zone, to the jungles of the equator; and it often merges into open country through parkland and savanna forest. Grassland includes all fertile unforested land, steppes, prairies and pampas, where vegetation of one kind or another is present throughout the year, and agriculture is possible. Deserts are regions where, owing to the scanty water supply, the life of plants and animals is either in abeyance or non-existent for at least the greater part of the year, and ordinary agriculture is impossible. Deserts are of two kinds: those which are *physically* dry, with little water present in any form, and those which are *physiologically* dry, where, though water may actually be abundant, it exists in a form inaccessible to living beings. Much of the Sahara and Gobi are deserts of the first type. Some salt deserts belong to the second category, but by far the largest physiological deserts are the frozen circumpolar belts where, for the greater part of the year, the water is locked up as snow and ice, and all life must exist under physiologically dry conditions. These types of country blend and grade into each other, not only at the limits of the greater formations, but also where local conditions change within the climatic formations themselves.

In some cases it is difficult to decide whether similar formations which do not lie adjacent have arisen independently under similar conditions, or whether their present likeness is due to past continuity. In the first place we may expect to find a difference, and in the second a resemblance between the flora and fauna. An excellent example of this occurs in the Sayansk Mountains at the source of the Yenisei River in Siberia. The spurs of this range, which form the northern bastions of the Central Asiatic highlands in lat. 52°–54°, are clothed on their northern slopes, four or five thousand feet above sea-level, with forests of the same type as the low-lying coniferous "taiga" between lat. 56° and 72°. Between the climatic forest formation and the mountains is an open steppe formation with a completely different flora. Now it is well known that altitude in warm climates may produce a state of things comparable to that found on low-lying ground in high latitudes— for instance the flora of high mountain tops has a general resemblance to that of the polar regions[1]. At first sight this seems to be the case in the Sayansk Mountains; but on investigation it appears more likely that, at an earlier epoch, coniferous forest

[1] Generally speaking, the mean temperature falls 1° F. for each degree of latitude north (or south) of the equator; and the same for every three hundred feet of altitude. The observations of the Everest expeditions suggest that certain species of birds, believed to breed only in Siberia, may yet be found to nest much further south in the Himalayan region and the adjacent highlands (4).

extended over the whole region, and at the end of the Ice Age, as the glaciers retreated northwards, the part of the forest on the mountain slopes was cut off from the rest, and the steppe flora crept into the intermediate zone with the altered conditions (35). Hence while altitude compensates for latitude as a general rule, its effect in this case has been to conserve an ancient formation, not to produce a new one.

My first expedition was in 1914 when I accompanied the late Miss Czaplicka to the cold desert tundras of Siberia, and after this lapse of time I can only marvel at the youthful optimism which led me without preliminary training, north-east in the footsteps of Middendorf, and at the good fortune which attended my ignorance. However I learned there the handicap of lack of preparation, and determined that next time opportunity offered a journey abroad, I should be better equipped. The occasion arrived three years later, during war service in south-west Russia. The Danubian steppe, though but the fringe of the great Eurasian grassland formation, is most impressive to the naturalist, and no doubt intensive study of the conditions in its seasonal and topographical semi-desert tracts would throw light on certain bionomical problems which are difficult to solve elsewhere. The contrast of tundra and steppe, their analogies and differences, first suggested the present studies, and when in 1922 the opportunity to travel abroad occurred again, I purposely chose the tropical jungle of British Guiana as a collecting ground, to complete a trilogy of Woodland, Grassland and Desert extremes. The manner of the proposed survey was more difficult to settle than the matter. Long ago I had fallen under the spell of *From North Pole to Equator*, and only those who have travelled through countries similar to those which Brehm visited can fully appreciate the insight and vigour of his descriptions of scenery and weather, and his unfailing gift for selecting the outstanding characteristics of animal and plant life in the wilderness. But, with all his charm, Brehm wrote for the nature lover more than for the naturalist, and his essays serve as a spring of happy reminiscence for the returned wanderer rather than as a guide to the œcologist on the spot. I began to wonder whether it was possible to offer not only picturesque descrip-

tions of climate and scenery to the traveller, but also to trace
the influence of these on the fauna with the greater precision
demanded by the student. For all travellers are not naturalists.
One I knew who travelled for a long September day across the
steppes, and, with eyes averted from broad horizons, ripe barley
fields, and hollows mist-blue with thistles and chicory, perused
the works of Mr Charles Garvice. And some students regard
the world merely as a collecting ground, or as a large open air
demonstration in laboratory science. I once saw a meteorologist
who, for the first time, watched the glory of midnight sunshine
on the snows of the arctic swamps and then remarked unmoved
that "the temperature was pleasantly high considering the
latitude and the season."

> Primroses by the river's brim
> Dicotyledons were to him,
> And they were nothing more.

It may be said at once that neither of the types of which these
individuals are representative will find anything to interest or
profit them in the following pages, which are but the outcome
of the happiness I myself have found in the curious and beautiful surroundings whither the study of natural history has
led me.

The aim to satisfy at once the nature lover and the naturalist
presents a double difficulty, for the manner must not be too
technical on the one hand, nor the matter too diffuse and
descriptive on the other. The first stumbling block I have
tried to avoid by relegating most of the scientific names to an
accessible but unobtrusive position in the index, and by trans-
ferring long references from the text to the sectional biblio-
graphies. In the present state of our knowledge the second
difficulty is the greater of the two, for little experimental work
has been done on bionomics, and the subject still rests largely
on generalisation and inference. But this very fact will be the
book's best justification, if thereby attention is drawn to the
rampart of our ignorance, as yet scarcely breached, and an
objective for another assault suggested.

The animal can be moulded to the environment either by

modification of form and function (structural) or by modification of behaviour and life history (bionomical), though the distinction between the two kinds of adjustment cannot always be finely drawn. On the whole, structural adaptations are determined chiefly by topographical factors, while bionomical adaptations are in the main related to climate. Indeed, figuratively speaking, the topography of the country and the variations of the climate represent the same part in the make-up of the environment as morphology and physiology play in the study of the organism. Thus the mammalian fauna of open treeless plains, tundras, steppes, pampas, high plateaux or tropical deserts, has, irrespective of climate, certain features in common. In each the dominant forms are mobile running animals or subterranean rodents of social habits; and the structural adaptations of the limbs to leaping or digging are often similar in different parts of the world. Naturally, species adapted to climbing are absent in the open country, but, where the plain or desert is broken by cliffs and ravines, it is frequently occupied by a secondary population whose members show affinities to the inhabitants of forests elsewhere, and, like them, are able to invade all three spatial dimensions. Thus typically the hyrax is a dweller in desert rocks, but certain African species inhabit hollow trees; and the wall-creeper of Central Europe, a cousin of the little woodpecker-like tree-creeper of our woodlands, lives in crevices in crags and precipices[1]. Again, animals which run upon a shifting or slippery surface are often provided with supporting flanges along the toes. This is seen in many desert-living reptiles, notably in one of the geckoes of the Caspian region, which, forsaking the climbing habits of its congeners, has lost the characteristic sucking finger discs, and developed lateral flanges to the digits. This structural adaptation finds its parallel in the "snow shoes" of the wood-grouse of the subarctic forests which pass the winter amid heavy snow. The lateral scales of the toes grow out to form a horny fringe which helps to support the bird in soft drifts or on slippery branches.

These are similar modifications of structure brought about

[1] The wall-creeper lacks the stiff supporting tail feathers of the tree-creeper.

by similarities of form in the environment; but, when habit and behaviour are considered, the determining factor is more often the climate. This is the case with the vast behaviour complexes connected with migration, hibernation, with breeding habits which are due to seasonal sexual activity, and to some extent with food-getting.

Some writers consider that, in the past, animal œcologists have laid too much stress on structural adaptations. For instance, Shelford (37) regards them as of secondary importance, and emphasises the necessity of regarding the animal kingdom from the point of view of reaction to environmental stimuli. But apart from the fascination of the study of morphological adaptation, which has held naturalists both before and since Darwin's day, structural adaptations play a part in habit and behaviour analogous to that of tools in the development of human culture (11). The anthropologist studies the evolution of the weapons, boats, pottery, etc., and the crafts, customs, and ceremonies with which they are linked, as a valuable source of knowledge as to the distribution, history, and present condition of the races of mankind. Until recently at all events, the "bow-drill" was still employed occasionally in English cutlery factories to bore the ivory handles of knives. In 1914, three months after watching this ancient device in use in a Sheffield workshop, the writer saw the same tool used by a nomad Dolgan on the Siberian tundra to drill holes in a plate of mammoth ivory. Now if in this instance the anthropologist had employed strictly œcological methods, he must have confined himself to remarking that these two workmen, in such different phases of culture, could both bore holes. He would have missed the fact that they performed the task by precisely the same means. In the same way, side by side with the environmental reactions of habit and behaviour, the zoologist must take into account the modification of organs which make such reactions possible.

This view is not in accord with those of certain modern œcologists. Shelford (37) compares "an African antelope running gracefully from a pack of hunting dogs, and an old-man kangaroo leaping from a pack of dingoes," and remarks that if the naturalist "notes mainly the specific peculiarities of the movements of the limbs and body of the pursued in each case, he will be dwelling upon specificities of little œcological significance, and missing the point of view of the œcologist altogether …it matters not if one animal progresses by somersaults so long as the two are in agreement in reaction to physical factors as indicated by the manner of spending the day, avoidance of forests, swamps, cold mountain tops, etc., entirely available to them."

But most naturalists will find the atmosphere of these œcological heights too rarefied for them; and therefore in the following pages, structure and behaviour will be considered side by side as mutually complementary to each other, though it must not be forgotten that the latter is more plastic than the former, and tends to evolve more rapidly.

A large class of structural adaptations are those correlated with the spatial level at which the animal lives, either under or on the earth, in the water, in trees or in the air. Sometimes such adaptations are common to a number of allied species or genera, and may even constitute isolated groups of considerable magnitude, such as the whales or bats. In other cases, a well-defined taxonomic order contains forms living at several levels. The Rodentia, for instance, include species which are cursorial, semi-aquatic, subterranean, arboreal and even aerial.

Another large class of structural adaptations is that concerned with obtaining food; and these may also be the diagnostic taxonomic character of a huge group occupying very diverse environments, as in the bugs, or may appear in a single genus, as in the crossbill, the skimmer, or in a more marked degree in the duck-billed platypus.

Adaptations to level are as a rule less restrictive than those concerned with food-getting, probably because the associated physiological changes are not so great. The bill of the snipe or woodcock is an instrument beautifully adapted for its purpose, but the normal range of these birds is correspondingly restricted to regions where the soil is soft enough for probing for food and contains abundant earthworms. On the other hand, the ant-eaters and porcupines are natural groups within which the members are adapted to life in various levels. The little arboreal *Tamandua* occupies a different stratum of the forest from the great fossorial *Myrmecophaga*, and yet their feeding requirements are similar, though very specialised. In this case the conquest of the tree-top level has been possible because ants and termites, which are the staple food of these anteaters, extend thither; and, as long as the food supply is assured, the structural difficulties in the way of arboreal life can be overcome.

With adaptations to level and food-getting should be included the vast class of adaptations for defence and offence, mimicry, obliterative coloration, etc., but as these will be discussed in a later chapter, they will be mentioned here only in passing.

The three great manifestations of animal activity are locomotion, feeding, and rearing offspring:

> Warum treibt sich das Volk so und schreit?
> Es will sich ernähren,
> Kinder zeugen, und die nähren so gut es vermag.

Thus we find that the third class of structural adaptations is that concerned with the rearing of young, but it is smaller than the others. Adaptations for the protection and development of offspring are of course innumerable, but the greater part are bionomical, or else affect the juvenile rather than the adult structure. The reason is obvious, for when the egg or embryo leaves the mother's body, organic continuity is lost and the offspring becomes a separate being which must work out its own salvation. All that the parent can do is to place it in the environment most favourable for development. This may be selected by the mother, as when the frog deposits her eggs in a pool and leaves them to hatch, or it may be specially constructed by her, as in the hunting wasps or the Australian mound-birds.

It must not be overlooked that some apparent "adaptations" are not modifications to environment at all, but are part of the animal's inheritance from less specialised ancestors. The South American "seed-snipes" are found in arid plains and plateaux from Peru to Patagonia. They are usually classed with the waders, but in appearance and behaviour they resemble the Gallinæ, and agree with the latter in their seed-eating habits and possession of a crop. At first sight these birds seem to be waders, adapted to life in barren semi-desert country where animal food is scarce; but there are grounds for believing that these characters are really primitive, and that the seed-snipes are the nearest living representatives of an ancient generalised stock, from which both the Gallinæ and modern waders are descended (25).

It follows that an adaptive structure does not necessarily function in perfect harmony with a particular environment. It may be under-adapted, in the sense that it is a comparatively recent acquisition which has not yet reached its optimum development under Natural Selection, or it may have evolved under different conditions, and been brought subsequently into a new environment. Wallace (44) describes a tree-kangaroo in the forests of New Guinea which is perhaps a case of this kind. This kangaroo, one of a family specialised for life in open plains, is arboreal in habits, but it is a slow and clumsy climber. However, modification seems already to have taken place in the direction of fitting it to life in its new surroundings, for the tail differs slightly in form and function from that of its saltatorial congeners. Certain writers also suggest "over-adaptation," but our knowledge is as yet too slight for such a conclusion. Some of the instances cited, like that of the kangaroo quoted above, presumably arise from change of environment; and others, for example the enormous development of the antlers of certain deer, though they may constitute a drain on the vitality and even a danger to the life of the organism, may have compensating advantages in other directions which we do not at present recognise. It appears safe to say, however, that an animal which has become specialised, either structurally or physiologically, gives a hostage to fortune whose value is in proportion to the degree of specialisation, for such an animal has imposed a limit to its own dispersal.

With structural adaptations must be included those that are purely physiological. So far these have been little studied in the higher terrestrial forms, and yet they are of great importance. Much useful work has been done, especially in America, towards grouping animals into "associations" and defining their characteristic environments; but this is preliminary, and the real task of œcology must be to determine why the animal is fitted to range only within these limits, not merely that it does so.

Even where the work has been undertaken it is not always possible to decide whether the organism has become physiologically adapted to the environment, or whether it is able to occupy such and such surroundings in virtue of some constitutional peculiarity which arose from other causes. Swain has shown that the Californian coyote, which inhabits dry waterless plains, has urine of high specific gravity (41). He suggests that the power of conserving the water supply may have been acquired in response to arid conditions, but it may be a constitutional peculiarity. The latter seems to be the case with reptiles (37) which excrete nitrogen as uric acid, a substance of low osmotic pressure which passes out in a dry state with the fæces. The water-storing capacity of the Australian toad *Chiroleptes platycephalus* is evidently an ancestral feature more or less common to all batrachians, but possibly increased in response to desert conditions.

We are sometimes tempted to ask what is the use of all this morphological and physiological adaptation, for many groups, without particular modification, seem able to maintain themselves successfully under the most diverse climatic conditions. In Siberia, the badger ranges from the sub-arctic forests to the sandy steppe deserts; and Merriam (28) has shown that the pocket gophers of North America flourish in every environment, from dry deserts to the moist sub-tropical forests. The reason that these and similar forms are able to range so widely is often because they are subterranean in habit and have a fairly catholic taste in food. In this way they are protected from the extremes of heat and cold, and from famine. Such animals do not adapt themselves to the environment: they try to evade it.

Other forms flourish in all kinds of environment by bionomical adaptation. Thus squirrels are found in forests from Labrador to Guiana, and can tolerate the winter conditions in the northern limit of their range through their capacity for hibernation.

The social bees offer a good example of bionomical adaptation. The humble-bees (Bombinæ) have their headquarters in the temperate zone. A few range into the high north of both Old and New Worlds, but the tropics are relatively poor in species of *Bombus*, and their place is taken by the Meliponinæ, a family of small stingless bees of social habits.

In Britain, the humble-bee's nest is founded in spring by a single fertilised queen who has hibernated through the winter. Her earliest progeny are females like herself, but smaller, and with the reproductive organs undeveloped. They are the worker caste who perform the duties of the nest. Later on in the summer, unfertilised eggs are laid which develop into males. These mate with the young functional females or queens and then perish. Of all the original colony, only the fecundated queens survive the winter. This is the normal cycle in temperate climates, and it is really determined by nutrition. The workers are females, but they appear when the queen's reproductive activity is high and while trophic conditions are poor; therefore they are small and their generative organs are not fully developed[1]. Even when the food supply improves, the population of the nest is so great that there is not enough to go round, and the summer eggs develop into workers also. Later on, when the number of workers is at its maximum, food is plentiful and the fecundity of the queen begins to diminish. For the first time there is a surplus of provisions, and the young larvæ are better nourished, and are able to develop sexually. The males are derived from parthenogenetic eggs laid by the queen, or apparently in some cases, by the workers.

In the tropics, where there is no winter to be tided over, humble-bee colonies are said to be perennial, and this is the case with most if not all the Meliponinæ. Food is available all the year round, and the reason that worker females appear is presumably because the nurseries are crowded and the larvæ are on short commons. To meet the resulting over-crowding, the Meliponinæ swarm just as the honey-bee does, that is to say

---

[1] Cf. Wheeler (47), p. 120. On the other hand, Sladen (39) considers that an abundant supply of food alone is not sufficient to make a worker larva develop into a queen. The food itself may be of different quality, or it may be that the queens arise from eggs laid late in the foundress's life.

one of the queens leaves the nest accompanied by some of the workers. The fact that the honey-bee cannot found a new colony without swarming is a strong argument for the tropical or sub-tropical origin of this species. There is however a difference between the swarm of the honey-bee and the Meliponid. In the former, it is the old queen who departs, and her daughter succeeds her in the hive. She leaves in state, surrounded by a cloud of workers who follow her with devotion. The natural history of the stingless bees is still little known, but von Ihering (43) finds that in this group the mature queen is unable to fly, owing to the enormous development of the abdomen full of eggs and the abraded condition of the wings. Hence it is the young queen which leaves the nest; and although a number of workers accompany her, they do not appear to pay her any particular attention, and according to the above-mentioned author, she is frequently found later wandering about all alone.

In the arctic zone we find another condition of affairs among humble-bees. Instead of the workers forming the largest caste in the community, they seem to be relatively rare, and the colonies themselves are small. During twenty years at Tromso in northern Norway, Sparre Schneider (15) "never saw a worker of *Bombus kirbyellus* and very few of *B. hyperboreus*."[1] Presumably the foundress and her progeny (or at least some of them) lay eggs which develop into males which mate with the young queens which appear in the early broods. An explanation which suggests itself is that not only is the arctic summer short but its onset is sudden. There is food only for a little while, but, while it is available, it is abundant, for the arctic flower fields in blossom are as rich as any meadows in the world. The foundress queen has less time and so lays fewer eggs, and the resulting larvæ are better nourished, so that functional males and females appear earlier than in the more protracted summer of the south.

Climate, soil, vegetation, etc., play an important part in en-

---

[1] This observation is not altogether borne out by those of others in the north. For instance, workers were collected in seven out of the eight species of *Bombus* recorded by the Canadian Alaskan Expedition of 1913–18.

vironment; but other influences must be taken into account also. The animal does not pursue an isolated path in time and space, affected only by such things as heat and cold, mountains and marshes; but each species as a whole, and each individual of the species, is more or less intimately connected with all the rest. Just as each stone of a handful thrown into a pond creates an ever-widening circle of ripples, which impinge upon the ripple circles of all the other pebbles, so every animal occupies the centre of what may be termed a bionomical complex, involving a number of other forms. Darwin's classical case of field mice and red clover may be taken in illustration. The mice have no direct influence on the clover, but the plant depends on the visits of humble-bees for its pollination, and mice destroy the bees' nests. Hence any increase in the numbers of field mice will presumably affect the clover crop adversely. If we compare a bionomical complex to a series of concentric circles, such an example as this will of course represent only a small section of arc of the innermost rings. The clover itself has enemies that devour it; the bees are attacked by other predatory and parasitic foes; and mice are the prey of cats, kestrels, and weasels. Such complexes frequently show interrelationships between most unexpected forms. Turner (42) has demonstrated one between redshanks and snipe on the one hand, and London omnibuses on the other. Before the development of motor traffic the rank marsh grass of the Norfolk Broads was regularly cut down and ground into chaff, which was sold to the London omnibus companies as horse fodder. In spring the grass mown in the preceding summer afforded the optimum breeding place for small wading birds. But when the horse disappeared from the streets there was no market for the grass, which consequently grew coarse and tussocky, and in some places gave way to long sedge. Small birds, such as snipe, peewits and redshanks cannot force their way through such thick growth, and their available breeding grounds were reduced. Other examples of associations between animals, and between animals and plants, will be discussed later on.

The limits of such associations can never be determined with our present knowledge of bionomics; but where the relationship is close, and the forms are structurally or bionomically adapted to one another, we can sometimes indicate at least the centre of the complex. Every degree of association is known, from cases where one animal is entirely dependent on another for its existence, to those where it figures only as a chance item on the menu of its enemies, or indirectly affects its neighbours by competing for food, breeding sites, etc. We are far from understanding even simple problems of this nature, and we are further still from the position of testing our ideas experimentally; but we are justified in holding that the way is worth preparation, because now and then it is possible to check our speculations by results in the field. Most of these cases relate to parasite and host, because this relationship is usually intimate and the environment of the parasite is well defined.

The subject of parasitism has of late years become a special branch of zoology, and is too wide to be more than touched upon here. In the strict sense of its Greek derivation, the word "parasite" means one who lives at the expense of, or eats the food of, another; but by use its meaning in biology has been extended to include cases where the host itself is devoured by the guest. Wheeler (46) has pointed out that the larval Ichneumonid which devours the tissues of the host caterpillar is thus not a parasite *sensu stricto*, for the Ichneumonid slays its prey as surely as, though more slowly than, the tiger. He proposes the term "parasitoid" for such cases, which are really a special and subtle modification of the predatory habit. But custom and wont are strong; and for most of us a parasite is still an animal which preys upon the living body of another in such a way that if death occurs, it is only by slow degrees.

Bionomically, the parasites of land animals fall into three classes, though there is of course a certain amount of overlapping[1].

I. Forms such as Tape-worms which live upon the food of the host and do not destroy it by direct attack, nor devour its tissues.

---

[1] Needless to say these three divisions do not represent stages in the evolution of parasitism, and Class III, the most important bionomically, does not necessarily include the most specialised parasites.

II. Forms which attack the host tissues, but do not bring about death unless they are present in large numbers. Such are many Protozoa, Round-worms, Flat-worms, some Arachnids, Insects (some Flies) and external parasites such as Lice, Fleas and Mites.

III. Forms whose normal cycle includes the death of the host, such as Flat-worms (in intermediate host), and Insects (some Flies and Hymenoptera).

There are many exceptions to Classes I and II, especially where the parasite has recently attacked, and not yet become properly adapted to, a new kind of host; but it is clear that on the whole these classes are of less definite bionomical importance than the third because their attack does not necessarily imply the destruction of the host. In fact, it is against the interests of the parasite itself to destroy the body that nourishes it: and if death supervenes it is either because the parasite is present in sufficient numbers to cause fatal exhaustion, or because bacteria secondarily invade the lesions in the tissues.

The third class includes a number of insects, chiefly belonging to the orders Diptera (Tachinidæ) and Hymenoptera (Fossores, Parasitica, etc.) which are parasitic on other invertebrates, the majority on insects, and are of great bionomic and economic interest.

Every species passes periodically through a critical stage, the weakest link in the chain of its existence, when the limits of its environment are specially straitened. Its increase and its range both depend on how it adapts itself to meet this crisis. The study of these critical times is important to the economic biologist in relation to the control of pests. Thus in Aphides the winter egg is provided with a tough coat, resistant to spraying and fumigation; and later in spring, when reproduction has begun, remedies are seldom radical, for some of the brood escape the douche and their fecundity is such that in a few days the infestation is as heavy as ever. The optimum time for applying preventative measures is after the hatching and during the development of the fundatrix, when she is unprotected by the eggshell, and before her destructive brood has appeared.

Malaria and yellow fever have been banished from many places because their mosquito carriers have been attacked at the vulnerable stage of their life history—when the aquatic larvæ and pupæ must come to the surface of the water to breathe and can be suffocated with oil.

Apart from the above, every species has its range limited by the presence or absence of certain essential factors in the environment. Sometimes it is a necessary food stuff. Liebig pointed out in his so-called "Law of Minimum" that the growth of a plant is controlled by the essential food stuff present in minimal quantity, no matter how abundant other food may be. This principle applies to animals also. Another limiting factor is the amount of suitable breeding ground available, for this really determines the range of a species. Thus the common frog spends the greater part of its life on land where it can adjust itself to considerable diversity of conditions, but its range is strictly limited by the presence of pools and streams in which it can lay its eggs.

Animals are generally intensely conservative in the selection of breeding environment, a fact which the experienced collector unconsciously recognises when he makes a "find" and offers no other explanation than that "it looked a likely place." Long practice has taught him to appreciate the complex of factors which influenced the animal's choice; for it must not be forgotten that the higher animal possesses the power of *choice* while the higher plant has *opportunity* only in the selection of its environment.

As a general rule, the breeding environment of the individual appears to approach closely to the ancestral environment of the species, and difference of breeding grounds in allied forms is sometimes accompanied by significant differences of appearance or habits.

Temminck's stint and the little stint nest on the tundras of the Lower Yenisei; but they are quite distinct in their behaviour and appearance in the field, and they select different environments for the nest, even within the same two furlongs. Temminck's stint haunts sandy ground near open water, preferably among willow scrub. The nest is lined with grass; the birds are wild and suspicious of approach; and, in their summer haunts at least, they seldom flock with other species. The little stint nests on boggy tundra, often at a distance from open water. If willows grow near, the leaves are used for nest lining. The birds are confiding and unconcerned at the approach of man, and frequently join mixed flocks of other waders. In fact the little stint, in appearance, behaviour, and choice of breeding environment, is a miniature edition of the congeneric dunlin, while Temminck's stint is the œcological representative on the tundra of the common sandpiper of our British lakes and streams.

The restriction of birds to particular environmental complexes is often very marked in mountainous regions. Chapman (7), who has studied the avifauna of the Andes, divides the mountain slopes into tropical, sub-tropical, temperate and "puna" zones. The puna is the Andean equivalent of the arctic tundra—harsh treeless wastes with a low temperature, lying between the forest and the snow line. He has shown that each of these "life zones" is inhabited by characteristic birds, which do not often range above or below them. For instance, in the tropical zone, are found some macaws which are distributed from Bolivia to Mexico, that is over 2500 miles of latitude; whereas, on the mountains, their range in altitude is less than a mile. The birds of the temperate and puna zones are very distinct. Those of the former—tanagers, parrots, flycatchers, and toucans—are all adapted to live in the dense forests which clothe this part of the mountain slopes, and are evidently relations of forms which live in tropical and sub-tropical forests on the rest of the South American continent. They have in fact sprung from the same stock; but, as Chapman points out (8), during their conquest of the mountain's temperate zone they have had to accommodate themselves to changes which would be as great as from, say, Ecuador to Ontario if expressed in terms of latitude instead of altitude, and consequently the birds have diverged considerably from the parent type. But the species characteristic of the arid heights of the puna—finches and oven-birds—are typical of plains elsewhere; and actually they appear to have descended from Argentine and Patagonian forms. During the evolution of their range, they have been subject to little environmental change, and consequently do not differ widely from their relations in the plains of temperate South America. Chapman's analysis brings out this distinction very clearly:

GENERA	PUNA ZONE (15,000–12,500 ft.)	TEMPERATE ZONE (12,500–9,000 ft.)
Generally distributed	27	11
Of south temperate zone origin	19	1
Of tropical zone origin	0	8
Restricted to "life zone"	7	28

The choice of environment has an important bearing on the segregation of species, physical or physiological. Physical segregation is brought about by geographical barriers, such as seas, rivers and mountains. The well-known faunistic anomalies of isolated peaks, islands, and lakes are often due to this cause. The curious distribution and differentiation of forms in the Galapagos Archipelago helped to turn Darwin's attention to the problem of the origin of species. Seventy per cent. of the birds are peculiar to the group; and each island seems to have an indigenous species or variety of lizard (*Tropidurus*), and probably of tortoises and other forms of life also [1]. The iguana *Cyclura ricordii*, after fifty years, was recently rediscovered on an island in a lake in San Domingo [29]. The rodent *Romerolagus nelsoni*, intermediate between hares and pikas, is confined to the slopes of Mount Popacatopetl in Mexico. A humming bird (*Oreotrochilus*) is found only on Mount Chimborazo in the Andes; and a mutational form of the tanager finch *Buarremon inornatus* is peculiar to the Chimbo Valley of Ecuador [8]. The Caspian Sea and Lake Baikal each possesses an indigenous seal; and the mountain lake Titicaca in South America contains the flightless grebe *Podiceps micropterus*.

Rivers frequently act as physical barriers. Preble [34] and Stone [40] consider that in the Mackenzie basin the caribou are divided into two or more races, and suggest that this is partly due to the river, which has separated the population into different herds, or associations of herds, which cannot mingle or interbreed, and have developed somewhat different habits.

Physical barriers may operate effectually on a smaller scale than those mentioned above. Every great climatic formation is broken into smaller local formations. Swamps and oases break the monotony of steppes and deserts; wooded tracts invade prairies; and forests contain natural or artificial clearings. The influence of these on the fauna depends on their extent, permanency, and degree of isolation, and naturally it is greatest where non-mobile or non-migratory animals are concerned. The determining factor is not always obvious. The kangaroo-rats (*Dipodomys*) are typical desert living forms, and are most abundant in the arid plains of south-west North America where they burrow in the sand. As might be expected, the important limiting factor in their range is water (rivers, lakes, swamps), but Grinnell (16), who has studied their distribution and œcology, has shown that it is not the only one and that temperature also plays a part. But anomalies of distribution still required explanation; for some species were found to flourish equally well in deserts and on moister river banks; and it has been suggested that the real determinant is not so much the dryness as the looseness of the soil in which the rodents burrow. Humid soil is usually heavy and the rats are poor diggers. Hence wetness in this instance turns out to be an indirect rather than a direct limiting factor in the range of the kangaroo-rats.

Less is known of physiological segregation than the subject seems to deserve. The controlling factor is generally food, and though our knowledge is still very limited, some interesting facts have come to light which have recently been summarised by Brues (5). He points out that where an insect feeds upon two species of plants, A and B, there is sometimes a tendency to split into an A-feeding and a B-feeding form whose paths henceforward lie apart and which may or may not show structural differences. This was indicated in the sixties by Walsh in America; and under the influence of the recently published *Origin of Species*, he suggested that "food races" might be species in the making, if the attraction of their native plant to each generation was sufficiently strong to segregate them from their relatives of the other race. This idea did not receive much attention at the time, but of late years it has been followed up experimentally.

Craighead (9) worked with various Longicorn beetles which feed on oak, hickory, pine and other woods; and his results suggest that it is possible gradually to isolate strains within the species which have a tendency to choose for oviposition wood of the kind in which they themselves were reared. Thus one form, which normally lives in pine, rejected spruce, but it was found that when the half-grown larvæ were transferred to spruce, the beetles when they emerged chose the latter wood in preference to pine.

Schroeder (36) experimented with another beetle whose larvæ feed upon and skeletonise the underside of the leaf of the willow *Salix fragilis*. He transferred some larvæ to *Salix viminalis* whose leaves are pubescent, and found that they soon learned to push the hairs aside to reach the parenchyma, and one even made a gallery through the tissues. After four generations, the beetles had thoroughly adopted the leaf-mining habit, and showed a growing preference for *Salix viminalis*, the percentage of females which selected it for oviposition rising from nine per cent. to forty-two per cent.

A Trypetid fly (*Rhagoletis pomonella*) in North America, which originally fed on some native plant, has now become a pest of cultivated apples. A smaller race, of somewhat different habits, is a pest of blueberry; and there is reason to believe that the apple and the blueberry flies are forms of the same species, although they never now interbreed and it is impossible to rear the apple form on blueberry and *vice versa* (31).

The adaptation to a new host plant may arise accidentally. *Aphis grossulariæ*, a common pest of currants in this country, is morphologically identical with *Aphis viburni* found in spring on the wild guelder-rose, and experiments suggest that although the *viburnum* form can establish itself on the currant, the descendants of such transferred forms cannot return to the guelder-rose. In this case the intermediate host is probably some other weed, and the currant is merely accidentally infected in the course of migration (18). The caterpillar of *Lasiocampa quercus* feeds on oak, and Pictet[33] transferred the larvæ to pine. Many individuals died, for their jaws were not wide enough to bite the pine-needles, but a few learned to nibble the tips of the needles to reach the parenchyma, and these survived. The second generation were found to adapt themselves to pine, and when given oak leaves, they attacked them likewise from the tip[1].

Sometimes the change to a new host plant is associated with definite structural changes. Marchal[27] took eggs from a peach tree Coccid, and transferred them to the *Robinia* tree. The resulting adults took on the size and appearance of another supposed species of scale-insect already known on the *Robinia*, and attempts to transfer them back to the peach tree failed.

Thus so far the experimental results are suggestive, but too limited to be conclusive.

Even difference in size, dependent upon nutrition, may "split" the species, if thereby mating is made physically impossible. The size of the Chalcid ectoparasites of a leaf-mining fly is determined by the size of the host larva at the time of oviposition. When the host is only half-grown, the parasite has less to eat, and the pupa either does not emerge or the resulting imago is very small. Another parasite of the same fly does not begin to feed until the host larva itself is full-grown, and the imagos do not show much disparity in size[19]. Pantel[30] and Keilin[22] have remarked on the dependence of dimensions on nutrition in certain parasitic Diptera; and some of the innumerable varieties described among the Braconid parasites of Aphididæ may have arisen in the same way.

It is well known that the numbers of many, perhaps all, species of animals fluctuate from year to year, according to environmental conditions. Thus if the early summer be cold and wet, it is safe to prophecy a scarcity of wasps later on, since the mortality among the queens will be high and few colonies will be founded. Precise data are scanty, but in this connection MacFarlane[26], Hewitt[20] and others have pointed out the value of the fur returns of the Hudson Bay Company which extend over the nineteenth century. These figures indicate only

---

[1] I have not been able to consult this paper in the original, and the above account is taken from Brues' work already cited.

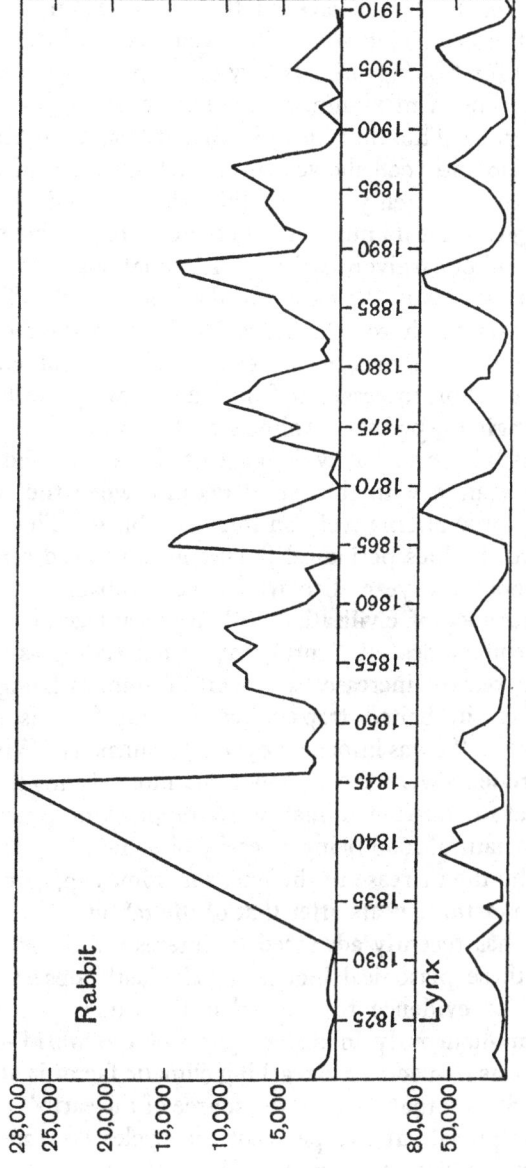

Fig. 1. Periodic fluctuations of the rabbit and lynx in Canada.

the numbers of animals captured, not those actually in existence; and moreover, being averages for the whole region, they take no account of local fluctuations, but even so they show that the numbers of fur-bearing animals vary over a period of years, and that the seasons of maximum abundance occur at more or less regular intervals. Thus the returns for the "fisher," a large marten, between 1840 and 1908 showed seven such maxima, which occurred about every ten years. As the fisher is omnivorous the figures suggest that its numbers fluctuated irrespective of food supply. In other carnivorous forms, the variations in the curves can be correlated with the curves of the food animals. Thus, in eight-six years the lynx, which preys chiefly on the American rabbit, showed nine maxima which tended to occur either in the years of rabbit increase, or from one to two years later, in cycles averaging 9·5 years. Animals such as the fox, wolf, and marten which have a fairly wide range of diet, do not show such a close correlation with the rabbit because when the latter is scarce they prey alternatively on mice and birds. The returns for the wolf, besides periodical fluctuations, showed a marked decrease about the year 1870 which was probably connected with the advance of civilisation and the disappearance of the bison. In some cases the figures may be misleading as regards the precise year of increase by reason of animals being more readily taken in baited traps when natural food is scarce. Preble (34) says: "I was informed by W. A. Burman of Winnipeg that small rodents were so common in Manitoba during the same autumn that fur-bearing animals were trapped with much difficulty, their natural food being so easily obtained." This partly explains why the increase in the lynx sometimes appears in the returns two or three years after that of the rabbit.

Elton (13) has recently advanced an interesting theory to account for these periodical increases. He justly observes that there is some evidence to show that the same species may increase simultaneously in distant parts of the world—a fact which suggests that some far-reaching climatic factor is at work. He also points out that the sun, the source of the earth's energy, varies in output about five per cent., in cycles corresponding with the appearance of sunspots. The cycles have an average of about eleven years, and certain climatic changes in the earth accompany the sunspot maxima and minima respectively. Among others, the average annual earth temperature at the sunspot minimum appears to be about o·6° C. greater than at the maximum, that is to say enough to shift the isothermal line for eighty miles. The effect, though liable to the interference of noncyclical factors such as volcanic eruptions, is greatest at the tropics and progressively less towards the poles. Elton attempts to show that increases in the numbers of different animals take place on an average of every ten or eleven years, and that these increases occur near the period of sunspot minima. The correlation of climate variability with solar cycle has not yet been worked out with precision; but rainfall, temperature, storm-tracks, etc., are probably involved, and the two former at any rate may determine the vigour of plant growth.

The statistics show a double cycle for animal life —one averaging about 3·5 years, and the other averag-

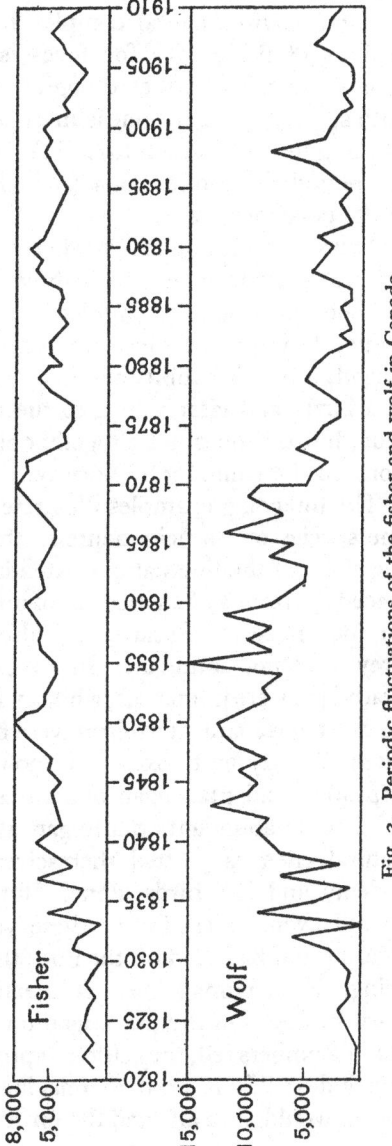

Fig. 2. Periodic fluctuations of the fisher and wolf in Canada.

ing about ten years. The latter may be correlated with sunspot cycle. The former is at present unexplained, and must be due to some environmental complex not yet recognised. Elton's analysis of the curves for foxes is specially interesting, and brings out the fact that though the two cycles are present in both species, the short one is more obvious in the arctic fox and the long in the common fox. This is what might be expected if the periodicity is due to a solar cycle whose effects are less felt in the polar regions.

At present it is not known whether the decrease in the number of these predatory animals is directly due to the disappearance of food, or whether diminished food supply may indirectly reduce fertility. MacFarlane states that, during the increase period, a female rabbit may have six, eight or even ten young at a birth, and later return to the normal number of three or four; but this does not altogether agree with Collett's observations on the lemming in Norway.

The foregoing examples illustrate the immediate influence of one species on its neighbours in the bionomical complex, but the effect of the fluctuations extends beyond this. Cabot (6) has traced the results of an increase of mice on the fauna of Labrador. In 1903 mice were scarce and also ptarmigan, while birds of prey were not abundant. In 1904 mice were more plentiful, ptarmigan were numerous, while owls and falcons had increased. In 1905 mice and ptarmigan were both very abundant, as were birds of prey and foxes. In 1906 the mice had almost disappeared, and ptarmigan and foxes were scarce. Thus when mice were abundant, ptarmigan also increased because their natural enemies glutted themselves with the easily captured rodents and left birds alone. But the web of this complex spread even wider, for in these years trout swarmed in the streams and battened on the mice that they caught while swimming. Even human life was affected, for the Indians rely in great measure on the ptarmigan for winter food, and when the birds' numbers fell, they suffered proportionately. Indeed Cabot concludes: "It is hard to imagine any other natural change which would have affected the fortunes of all the other creatures ...from man to fish as did the coming and going of the mice during the years from 1903 to 1906. Only fire could have done the like."

This summary of the scope of bionomics indicates how wide is the field, and that zoology must call in the sister sciences to cultivate it.

Firstly, it is necessary to study the topography and climate of the region. Topography includes the soil and its origin, the general configuration of the country, whether mountainous or low lying, dry or marshy, the number and courses of the rivers and streams, whether wooded or open, and the kind of vegetation composing the different formations. The study of the climate is most important. The two principal factors are temperature and rainfall, and here not only the annual or monthly minima and maxima, but the daily rise and fall should be considered. The physical contours of the country must be taken into account with the temperature readings. In stations among or near hills, the winter records are usually taken in valley bottoms; and since damp cold air tends, as it were, to drain into the valleys, the readings frequently give a lower figure than the real mean of the country (23).

With the rainfall, it is necessary to consider not only the mean monthly fall, but also the character of the rain itself—whether it comes on in violent storms and so runs quickly from the surface, or whether it is a gentle precipitation which soaks in. In addition the amount of humidity in the atmosphere and the amount of cloud should be considered. Humidity must always be taken into account in relation to temperature. The absolute humidity percentage of warmer or colder air may be the same; but the relative humidity of colder air is less (to a warm-blooded animal at any rate) because the air on coming into contact with the body is warmed, and its relative humidity percentage falls. The absolute mean humidity percentage in winter in England is about the same as that of Tobolsk in Siberia—that is to say about eighty per cent.; but in January the mean temperature in Tobolsk is − 2° F. as compared with 39° F. in England. So that by the time the surrounding air is heated by the living body (approximately to 60°) the relative humidity of the Siberian air is about one-fifth that of the English. In other words, cold air is physiologically dry (23).

Secondly, the distribution and life history of the species to be studied should be known as thoroughly as possible. Before beginning an œcological investigation, we ought to know the duration of embryonic life, and of post-embryonic development and growth, and the comparative duration of adult life.

Thirdly physiology, with its ancillary biochemistry, must be called in to determine by experiment the physiological reactions of the animal to physical factors such as heat, cold, evaporation, CO_2 content, light, shade, etc.

Fourthly, the student must know something of systematic work. This is popularly supposed to be a dry-as-dust branch of zoology. In fact, the systematist may be called the dustman of biology, for he performs a laborious and frequently thankless task for his fellows, and yet it is one which is essential for their well-being and progress. In studying the bionomics of a particular species, it is necessary to be acquainted not only with the type, but with its variations; and if the work seems dull and profitless, let the student remember that some of the most striking discoveries in bionomics have been rooted in systematics. Bower (3) has recently written:

In my view, no one is properly fitted to investigate questions of evolution unless he should have acquired by practical study of some natural group of organisms sufficient experience of formal morphology to classify them himself, or to criticise the classification already advanced by others.

Lastly, it cannot be too often insisted that even the most trivial records may prove of value ultimately. Nothing is too small to note down. Museums are filled with specimens which are often worse than useless because no details have been kept of the conditions of capture. The slightest variation may make all the difference in the œcology of a species. Dahl (10) describes two doves in the Bismarck Archipelago which both feed on the same fruit; but one species invariably feeds in the branches and the other picks up the food from the ground. This habit is so pronounced that, even when captive in the same cage, the birds exhibit the same difference in behaviour; and, for aught

we know, this may have survival value in the struggle for existence, where the ultimate success or failure of the race often
depends on factors apparently small or obscure. A Scarabæid
beetle, *Ochrosidia immaculata*, became very abundant in a certain
area in America. Its larvæ were parasitised by a wasp, and, as
the numbers of the beetle declined, those of the parasite sank
also. Meanwhile a related Scarabæid, *Popillia japonica*, whose
habits and life history are very similar to those of *Ochrosidia*,
increased greatly in numbers; but although the parasite population was diminishing through lack of food, only two successful
cases of parasitisation of *Popillia* were recorded. The *Ochrosidia*
larvæ are somewhat larger than those of *Popillia*, and at first
sight it seems as if size might be the determining factor. But
observations showed that *Ochrosidia* larvæ of all ages were liable
to attack, and that the larvæ of the second instar were actually
smaller than those of *Popillia* larvæ of the third instar which
lived beside them. The occasional parasitisation of *Popillia*
shows that the wasp larvæ can actually feed upon either species;
and what really determined that *Ochrosidia* should be chosen
and *Popillia* ignored is still unknown [21].

The following case may be cited in support of the importance
of the minute record. In South Russia in 1917 I noticed that
the house sparrow was the only sparrow found round dwellings
in the neighbourhood of Odessa, although the tree sparrow
occurred in trees along the cliffs [17]. But westwards across the
Dniester, the tree sparrow was as common as the house sparrow,
and lived side by side with it around buildings and rickyards,
until in the country near Galatz in Bessarabia it out-numbered
the house sparrow by two to one, and was the domesticated
"sparrow" of war camps and villages. Unluckily, circumstances
did not allow of a census in the transition area; and I have re-
gretted this the more in finding subsequently how conflicting
are the accounts of the tree sparrow in the region, and how
anomalous is its distribution elsewhere. In 1836, Drummond [12]
found that the tree sparrow completely replaced the house
sparrow in Bulgaria and parts of Serbia; but Elwes and
Buckley [14] in 1863 did not observe the species in towns along
the Black Sea, though flocks were seen in the fields of Mace-

donia. Kennedy (24) also found that the house sparrow, and not the tree sparrow, occupied the town of Novorossick. In Great Britain, though widely distributed, the tree sparrow is local and does not replace the house sparrow; but in Ireland, which it has invaded apparently since 1850, its distribution offers some curious problems. For instance, it seems to be the only sparrow round the houses of the island of Inishtrahull off the coast of Donegal (32), whereas the neighbouring Rathlin Island (2), and probably Tory Island and Aranmore possess only the house sparrow, which is also the dominant form on the adjacent mainland.

Facts of this kind, involving changes in the range and behaviour of closely allied forms, bring home to us the amount of work that remains to be done before we can isolate the environmental factor or factors which determine, not only the distribution, but, as in the case of the Bessarabian sparrows, a change of habits in a common bird. Moreover they emphasise how irreconcilable are the classifications of taxonomy and œcology; for in Odessa the œcologist would refer the tree sparrow to a different association from that of the house sparrow, while at Galatz, little over a hundred miles away, it not only enters the house sparrow association, but even replaces its congener. In fact, from an œcological point of view, the tree sparrows of Odessa and Galatz must needs be regarded as separate species.

The position of bionomics in zoology is something like that of economics in the study of human life; and it has the same fascination and perils. "Il n'y a rien qui s'arrange plus facilement que les faits." There is the same satisfaction in tracing cause and effect in the study of what is *done* rather than of what *is*; but there is also the same temptation to select facts to build theories and support conclusions which cannot always be tested by experiment. The almost passionate devotion of certain œcologists to the terminology of their science is no doubt partly due to the demands of a study of recent and surprising growth; but it is also perhaps an unconscious acknowledgment that the conceptions underlying their definitions are too fluid to be covered by a formula be it never so precise, and too elusive to be fitted to a rule, however elastic.

BIBLIOGRAPHY

(1) BAUR, G. (1891). "On the Origin of the Galapagos Islands." *Amer. Nat.* vol. xxv.

(2) BEST, M. G. S. and HAVILAND, MAUD D. (1914). "Bird Migration in Rathlin Island." *Irish Naturalist*, vol. xxiii, No. 1.

(3) BOWER, F. O. (1923). "Remarks on the present outlook on Descent." *Proc. Roy. Soc. Edin.* vol. xliv.

(4) BRUCE, C. G. (1923). "The Assault on Mount Everest." London.

(5) BRUES, CHARLES T. (1924). "The Specificity of Food-plants in the evolution of Phytophagous Insects." *Amer. Nat.* vol. lviii, No. 655.

(6) CABOT, WILLIAM B. (1912). "In Northern Labrador." Boston.

(7) CHAPMAN, FRANK M. (1921). "Bird life in the Urubamba Valley of Peru." *Bull.* 117, *U.S. Nat. Mus.*

(8) —— (1924). "The Andes." *Natural History*, vol. xxiv, No. 4.

(9) CRAIGHEAD, F. C. (1921). "The Hopkins Host Selection Principle as related to certain Cerambycid Beetles." *Journ. Agric. Res., Wash.* vol. xxii, No. 4.

(10) DAHL, F. (1903). "Winke für ein wissenschaftlichen Sammeln von Thiere." *S.B. d. Gesell. Naturforsch. Freunde zu Berlin.*

(11) DARBISHIRE, A. D. (1917). "An Introduction to a Biology and other papers." London.

(12) DRUMMOND, H. M. (1846). "List of the Birds observed in winter in Macedonia." *Ann. Mag. Nat. Hist.* vol. xviii.

(13) ELTON, C. S. (1924). "Periodic Fluctuations in the Numbers of Animals: Their Causes and Effects." *Brit. Journ. Exp. Biol.* vol. ii.

(14) ELWES, H. J. and BUCKLEY, C. E. (1870). "The Birds of Turkey." *Ibis*, vol. vi.

(15) FRIESE, H. D. (1902). "Die arktischen Hymenopteren," in Römer and Schaudinn's *Fauna Arctica*, vol. ii. Jena.

(16) GRINNELL, JOSEPH (1922). "A Geographical Study of the Kangaroo-rats of California." *Univ. California Pub. Zool.* vol. xxiv, No. 1.

(17) HAVILAND, MAUD D. (1918). "Notes on some Birds of the Bessarabian Steppe." *Ibis*, vol. iv.

(18) —— (1919). "The Bionomics of *Aphis grossulariæ* and *Aphis viburni*." *Proc. Camb. Phil. Soc.* vol. xix, Pt. 5.

(19) —— (1922). "On the Larval Development of *Dacnusa areolaris*, etc." *Parasitology*, vol. xiv, No. 3.

(20) HEWITT, C. GORDON (1921). "The Conservation of the Wild Life of Canada." New York.

(21) JAYNES, H. A. and GARDINER, T. R. (1924). "Selective Parasitism by *Tiphia* sp." *Journ. Econ. Ent. Concord.* vol. XVII, No. 3.

(22) KEILIN, D. (1915). "Recherches sur les larves des Diptères cycloraphes." *Bull. Sci. Fr. Belg.*, 7ième série, vol. LXXVII.

(23) KENDREW, W. G. (1922). "The Climates of the Continents." Oxford.

(24) KENNEDY, J. N. (1921). "Notes on Birds in South Russia." *Ibis*, vol. III.

(25) LOWE, P. R. (1923). "Notes on the Systematic Position of *Ortyxelus*." *Ibis*, vol. V.

(26) MACFARLANE, G. (1905). "Mammals of the North-West Territories."

(27) MARCHAL, P. (1908). "Le Lecanium du Robinia." *C.R. Soc. Biol.* t. XLV.

(28) MERRIAM, C. H. (1895). "Monographic Revision of the Pocket Gophers." *U.S. Dept. Agric. N.A. Fauna*, No. 8.

(29) NOBLE, C. KINGSLEY (1923). "Tracking the Rhinoceros Iguana in San Domingo." *Natural History*, vol. XXIII, No. 6.

(30) PANTEL, J. (1912). "Recherches sur les Diptères à larves entomobies." *La Cellule*, t. XXVI.

(31) PATCH, E. M. and WOODS, W. C. (1922). "The Blueberry Maggot." *Maine Agric. Exp. Stat., Bull.* 308.

(32) PATTEN, C. J. (1913). "Discovery of Tree-Sparrows on Inishtrahull Island." *Brit. Birds*, vol. VII, No. 2.

(33) PICTET, A. (1911). "Un nouvel exemple de l'hérédité des caractères acquis." *Mém. Soc. Phys. Nat. Genève*, vol. XXXV.

(34) PREBLE, EDWARD A. (1908). "A Biological Investigation of the Athabasca-Mackenzie Region." *N. Amer. Fauna, U.S. Dept. Agric.* No. 27, *Biol. Survey*.

(35) PRICE, M. P. and SIMPSON, N. D. (1913). "An Account of the Plants collected on the Carruthers-Miller-Price Expedition through North-West Mongolia." *Journ. Linn. Soc. (Botany)*, vol. XLI.

(36) SCHROEDER, K. (1903). "Ueber experimentell erzielte Instinktvariationen." *Verh. deutsch. zool. Gesell.*

(37) SHELFORD, VICTOR E. (1911). "Physiological Animal Geography." *Journ. Morph.* vol. XXIII.

(38) —— (1912). "Aspects of Physiological Succession." *Biol. Bull.* vol. XXIII.

(39) SLADEN, F. W. L. (1912). "The Humble Bee." London.

(40) STONE, A. J. (1900). "Some Results of a Natural History Journey, etc." *Bull. Amer. Mus. Nat. Hist.* vol. XIII.

(41) SWAIN, R. E. (1905). "Some notable constituents of the urine of the Coyote." *Amer. Journ. Phys.* vol. XIII, No. 1.

(42) TURNER, E. L. (1921–22). "The Status of Birds in Broadland." Presidential Address to the Norfolk and Norwich Naturalists' Association.

(43) VON IHERING, H. (1904). "Biologie der stachellosen Honigbienen Brasiliens." *Zool. Jahrb. Syst.* vol. XIX.

(44) WALLACE, A. R. (1869). "The Malay Archipelago." London.

(45) WARMING, E. and VAHL, M. (1909). "The Oecology of Plants." Oxford.

(46) WHEELER, W. M. (1919). "The Parasitic Aculeata: a Study in Evolution." *Proc. Amer. Philos. Soc.* vol. LVII.

(47) —— (1922). "Social Life among the Insects." London.

PART I

THE RAIN-FOREST

CHAPTER I

LOW-LYING REGIONS, close to the equator, where the annual rainfall exceeds eighty or ninety inches, are usually clothed with a dense growth of jungle called by œcologists "rain-forest." Here vegetation perhaps reaches its culmination in the size and variety of its forms.

Typical rain-forest is found: in South America, around the basins of the Amazons and Orinoco, and in parts of the Central American Isthmus; in Central and Western Africa, along the courses of the Congo, Niger, and Zambesi rivers; in Madagascar; in the Indo-Malay States, Borneo, New Guinea, etc.

The particular tract of forest which is the subject of this and the three following chapters lies along the Essequibo River in Guiana, and is the northern coastal fringe of the rain-forest of the Amazons.

The determining conditions of this kind of forest are a high even temperature and abundant moisture. The mean seasonal variation of the temperature close to the equator at sea-level is seldom more than 5° F.; that is to say, the difference between summer and winter is less than that between day and night. In the Guiana forest the thermometer usually registers about 90° by day, sinking at night to 70° or thereabouts. In direct sunshine the heat is of course intense, and life is only tolerable because the boundaries between light and shade are sharply defined. It is possible to sit under the trees in a temperature which is pleasantly warm, and thrust the hand into a blistering glare where the mercury climbs to 150° or more. Except in the very deepest jungle, some sunlight filters to the forest floor through chinks in the foliage canopy overhead. These shafts of light strike as bright and hot as the rays from a burning glass; and as they move over the leaves, small invertebrate life creeps out of their way. I once watched such a sunbeam searchlight overtake a party of ants who were devouring the decomposing body of a caterpillar. Within half a minute they fled helter-skelter, and the remains of their meal shrivelled up as if before a furnace.

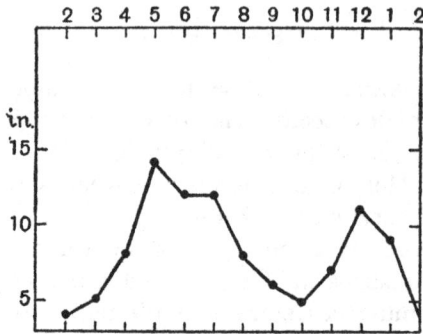

Fig. 3. Average monthly rainfall of rain-forest at Kartabo, British Guiana.

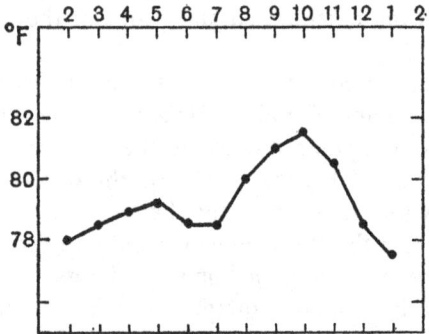

Fig. 4. Average monthly temperature of rain-forest at Kartabo, British Guiana.

For the most luxuriant development of tropical forest the rainfall must not only be copious—at least 80 inches annually—but it must also be distributed evenly over the twelve months. The rainfall reading alone is not necessarily an index to the vegetation. Chapin remarks (18) that in Sierra Leone, with 170 inches of rain combined with a four months' dry season, the forest is interrupted by grassland, whereas along the Upper Congo, where the rainfall is only 65 inches per annum, the forest is dense and unbroken, because the precipitation, though less in total amount, is regular.

The alternation of "wet" and "dry" seasons varies in different places. For example, in Guiana there are two rainy seasons in the year, each of two or three months' duration; but even in the so-called dry period there are occasional downpours, and prolonged drought is unknown.

Correlated with the distribution of the rainfall is the humidity, or quantity of aqueous vapour present in the air. In rain-forest this is always high, and sometimes approaches saturation. At sunset the temperature drops rapidly and a heavy dew often falls. Even in the daytime, especially after a shower, it is usual to see faint clouds of grey vapour floating away among the tree tops like puffs of steam.

The force of the downpour is another factor in the œcology of the forest. In the wet season thunderstorms of great violence are frequent, and the rain descends with a suddenness and volume unknown outside the tropics. The sun is shining, the forest glitters with a million lights, birds are on the move, and insects hum and dance from leaf to leaf. All at once a shadow is drawn over the sun, and all activity of bird and beast ceases as the sound of rushing rain rapidly approaches. An avalanche of water then crashes down, blotting out surrounding objects, and, as it seems, sweeping the very breath from the nostrils, bewildering and benumbing the senses. Every twig and leaf is bent and battered, and in a few seconds streams pour down the paths and the world seems changed into a thundering cateract. Then, as suddenly as it came, the storm passes, and the sun blazes out again before the roar of the storm sweeping over the tree-tops has died away in the distance. Even before the leaves have ceased to drip, or the land-crabs, tempted forth by the teeming water, have scuttled to covert again, the life of the forest is resumed. It is almost incredible how some fragile forms escape destruction under such terrific bombardments; and yet great butterflies, with a wing span of four or five inches and so delicate that the least touch marks them, emerge from the deluge unblemished. Nevertheless certain adaptations have arisen to meet these conditions. Some of the inhabitants of the "reservoir" plants are provided with suckers which enable them to cling to the leaves when their home is brimming over. Belt(10) remarks that many forest birds—trogons, motmots, parrots and toucans—breed in holes in trees, and he supposes that this offers some protection from marauding ants; but it is equally possible that in such situations the eggs and young are sheltered from rain. Occasionally these storms are accompanied by wind, which passes as quickly as it rises, and sends a volley of dead branches, ants' nests, and fruit, hurtling to the forest floor; but on the whole wind is not a factor of great œcological importance in the rain-forest, and the winds of the coastal regions at any rate are regular and steady, rising and falling at particular hours each day. In the early morning especially, it is often so still in the depths of the forest that a photographic plate can be exposed for an hour at a time without fear of a leaf stirring to mar the picture. Rodway(39) remarks that there are hardly any wind-fertilised trees in the forests of Guiana.

The seasonal change in the amount of light and darkness, which is such an important factor in the œcology of arctic regions, scarcely affects the life of the tropics. On the equator, the sun rises and sets about six o'clock all the year round, and throughout the rain-forest area the variation is not more than an hour or two either way. It might be thought that as the darkness is longer in the tropics than in temperate regions, the rain-forest would contain a proportionately large number of wholly nocturnal animals, but, although exact figures are not available, this does not seem to be the case.

There is a general idea that the tropical forest is the home of eternal summer, and that it is comparatively untouched by the passing of the seasons. This is not strictly true, for although excessive contrast between summer and winter climate and landscape are lacking, there is an orderly sequence of phenomena throughout the year. Not every kind of plant or animal is blossoming or breeding at the same time in the same locality. Many forms can find the optimum conditions only in the wet or dry seasons as the case may be. There is no general leaf-fall, but each tree sheds its foliage imperceptibly as the new leaves grow. Now and then a naked tree stands out among its better-clad fellows, but this only means that its individual "winter" has come on rather more rapidly than usual. However, at the close of the rainy season many trees put forth an extra abundance of young leaves; and for a week or two the Guiana forest takes on the most exquisite tints of bronze and yellow and fresh green, comparable to the autumn colours of the woods of Europe except that they are due to rejuvenated and not to withering foliage. In the same way, many insects, some butterflies for example, occur on the wing in numbers only during the wet or dry season and then disappear for some weeks or months.

The rain-forest is no more homogeneous in space than in time. Distinct associations of vegetation are discernible, and in various parts of the world the forest wears a slightly different aspect. Lang (27) has contrasted the Congo rain-forest with that of Guiana, and considers that the total insolation of the former is higher. In the African jungle rain falls heavily for a few hours, often at night, and sunny intervals are relatively frequent, even in the wet season. Moreover the giant trees are more scattered, and tower high above the general level of the forest, an arrangement which increases the surface of foliage exposed to sun and wind. In South America, rain sometimes falls in a steady drizzle for many hours, and in the wet season the sun may be hidden for days together. Seen from above also the forest is more uniform in height, presenting a flat expanse of foliage to the sky; and below, its aspect is darker and gloomier, and the growth of epiphytes among the upper branches is more dense.

The œcological grouping of the rich and diversified flora into associations and so forth is a task beyond the scope of this book; but a short time spent in the rain-forest soon establishes the fact that from the point of view of the fauna, the floral environment falls into a series of groups. Whether these possess any positive value, apart from the convenience of their recognition to the collector, is a matter of opinion; but there is no doubt that they facilitate his work. Chapin (18) was "early impressed by the general agreement in dispersal of the bird communities with corresponding plant associations," and the same holds perhaps to an even greater degree for the insect life.

The "associations" described here therefore are not to be taken as part of an established classification. The small sector of the Guiana forest to which these remarks apply is only a fraction of an enormous woodland system whose thorough exploration would certainly produce other and more comprehensive schemes. That given here represents only the working zones that I myself came to recognise when collecting in a small area; and it was encouraging to find subsequently that in the main my conclusions agree with those of Lang. He divides the Guiana forest as follows:

(1) Intermittently inundated forest.

(2) Higher lying forest.

(3) Secondary growth.

For the present purpose, the first division is further separated into (a) Swamp jungle, (b) Palm swamp and (c) Mangrove swamp.

The best and most comprehensive view of the rain-forest is obtained from open water, for when the approach is over cultivated ground, cumbered with the rank secondary growth of the tropics, the eye is distracted, and fails to grasp at once the size and extent of the primæval jungle. If the approach is by boat up one of the great rivers, which are still the only highways through the greater part of the forest region, the sight is one of unforgettable grandeur. On either side the banks are veiled by a wall of green foliage between one and two hundred feet high, towering above its own inverted image in the water. Here and there its splendid sameness is broken by a patch of coloured blossoms. The branches of the scarlet "rose of the forest" are thrust out over the river, and sprays of *Bignonia* and other flowering creepers, yellow, purple and red, hang over the trees. Seen from below the effect is that of some stately city street decked out for pageantry. This profusion of flowering climbers, which in some places hides the outlines of the trees themselves, is characteristic of the South American forest. The creepers cover the whole roof of the forest as with a canopy, and fall to its foot at the water side like a curtain. In fact the forests of the whole Amazonian region may be compared to a series of tables with many legs, separated by waterways, and each spread with a cloth which dips to the ground on every side. The table legs are the upstanding trunks of the trees: the cloths are the tangle of vines and lianas which cover them with a close network. This mass of creepers is not altogether the suffocating burden or host of parasites that it appears to be. In exchange for support, it affords shade which is essential to the well-being of the forest; and it has been shown that when the veil has been torn aside so that the sun can beat down on the roots, the giant trees perish. For this reason an artificial clearing is usually fringed with dead trees.

Here and there dark caverns yawn in the wall of foliage at the waterside. These are the mouths of creeks and streams, shut in by overarching branches from which long aerial roots hang down like stalactites. To enter these caves by boat is like passing from the open air into a vast dim hall, supported by immense columns. The trunks of the trees rise up for seventy or eighty feet without a branch, and the undergrowth is thin and straggling. The ground is strewn with dead leaves, though it may be remarked that the accumulation of leaf-mould is not very great, owing to the rapidity of bacterial action.

I knew this zone of damp forest beside the rivers as "swamp jungle." It is periodically submerged, and the soil is wet and marshy. Swamp jungle of this description sometimes extends far inland from the rivers. At certain seasons the flooded forest of the Amazons or Orinoco may cover some hundreds of square miles. This zone never appeared to be so richly furnished with lianas and climbers as the dryer forest, and I attributed their absence to the frequent inundations. Aloft they are scarcely missed, for their place is taken by a host of epiphytic and parasitic plants which crowd the branches, and the comparative bareness of the trunks of the great trees, like the unadorned pillars of a vast cathedral, only adds to the impressiveness of the scene. Here and there is a clump of coarse-growing herbaceous plants, such as *Heliconia*, with their gaudy scarlet and yellow bracts, but otherwise there is little vegetation under foot, and this perhaps accounts for the impression that the roots make upon the observer. Indeed the most arresting sight in the swamp jungle is the monstrous growth of the roots. In forests elsewhere, the imagination is instantly captured by the trunk and branches of the tree, and accepts them as the vital parts. The roots, if noticed, seem to be merely pedestals to hold the aerial parts upright, and even higher perception regards them as little more than an elaborate system of conduits. But in the tropical forest all this is changed, and the roots dominate the eye almost to the exclusion of trunk and foliage. The rest of the tree appears simply as a great lung which draws power from the sunshine to maintain the titanic battle that is waged over and under the soil. The slopes of a volcano with twisting lava streams, or Ragnar Lodbrog's serpent pit lignified into immobility, alone can give an idea of the intensity of the struggle of the roots. Here they stream above ground, and checked in their course rear up in slabs like tombstones in a cemetery, or bunch into knots like the arms of the Kraken. There they twist and grapple together before diving once more below the surface in the race for an inch of space or a drop of water. The scene may be static enough, but it is impossible not to describe it in terms of movement. Along the water's edge, where the soil has been sucked away, subterranean conditions are laid bare. Here the roots are woven into a mass so intricate that a knife blade can scarcely pass between them, and not one is undistorted by pressure. If five years or even five months of growth could be speeded into the space of a minute, what a spectacle of dynamic forces would be there.

Some of the great forest trees—certain of the Moraceæ for example—when growing in marshy places throw out buttress roots. These are flattened plank-like flanges which rise out of the ground and clasp the trunk sometimes for a height of twenty feet or more. The buttresses are frequently of monstrous or bizarre growth, twisted and convoluted to form clefts and caverns in which water and humus collect. Sometimes the whole tree turns in a half-spiral, and the great root folds are wrapped round it as sculptured draperies follow the lines of a statue. These root outworks often increase the basal girth of the tree

PLATE I

RAIN-FOREST WITH MANGROVE SWAMP

to such an extent that the bole, where it rises above them, seems slender out of all proportion, and yet measurement will show it to be four or five feet in diameter.

The fauna of the swamp jungle is at first somewhat disappointing to the collector, for the large or conspicuous forms, usually associated with the tropical forest, are for the most part absent.

The only bird song characteristic of the ground level of this region is the triple note of the gold-bird whose loud call, "Pi—pi—yo" repeated at frequent intervals, sounds almost sacrilegiously glad and careless in that silent place. The shade is too dense for any but small and unobtrusive forms of insect life, and mammals in the forest are generally shy and seldom seen. The periodical inundations naturally influence the character of the fauna. Where the great rivers may rise as much as forty feet, and forest tracts as large as some European countries stand for weeks under water, the most flourishing and widely ranging mammals are those which are either arboreal or amphibious. The former will be dealt with in more detail later on. As regards the latter, it is noteworthy that the Marsupialia, Rodentia, Carnivora and Ungulata have all amphibious representatives, viz. the water-opossum (*Chironectes*), the capybara and labba, an otter (*Pteronura*) and the tapir. In addition there is the wholly aquatic manatee. With the exception perhaps of the armadillos and the great ant-eater the other non-arboreal and non-amphibious mammals of the neotropical rain-forest are cursorial forms such as peccaries and deer which are able to escape rapidly from the rising water. The swamp jungle is also the haunt of several amphibious snakes, such as the water-boa, and it is the chosen home of batrachians with sucking finger discs which allow them to climb above the flood. There is the giant *Hyla maxima* whose sonorous croaking at night replaces the song of the gold-bird by day; and the delicate little *Phyllomedusa* whose weight scarcely causes the bending of the leaf to which it clings. There is also the curious *Hyla boans*, whose body when at rest resembles a yellow leaf in form and colour, and which, chamæleon-like, can change through the whole gamut of tints from brown to creamy white, while the spots on the flanks imitate a patch of blister on the leaf.

It is perhaps somewhat misleading to call the second division of inundated forest "palm swamp" because palms of different kinds are found singly or in groups everywhere in the forest whether the soil is wet or dry. There are however certain marshy places at the heads of creeks or beside the rivers where natural or artificial clearings exist, and these are frequently occupied by a "pure culture" of palms. Owing to the greater amount of light and the change in vegetation in such places, the insect fauna is in some respects distinct from that of the forest around. This distinction is not altogether one of kind. The differences of environment all over the world are frequently only of degree, that is to say that though the naturalist knows well enough that certain species will be abundant in one environmental complex and scarce in another, he may be hard put to it to name even one which is not common to both. He can only judge by averages, for immigrants from neighbouring areas wander in, and although their tenure may be temporary or precarious, their occurrence obscures that clear-cut division into "associations" or "communities," whose preservation simplifies the task of the œcologist. Thus memories of palm swamp conjure up visions of the great blue butterfly *Morpho menelaus*, and of the big green and primrose bees which swing by their mandibles from the fronds, but further reflection recalls that the former was even more abundant in the higher forest, and that the latter were at least equally common elsewhere. On the other hand, there is a certain red and black frog-hopper which I never took anywhere but in palm swamp, though two of its congeners in Guiana inhabited shady forest and open clearings respectively.

Palm swamp is in its way as impressive as swamp jungle, though with less austerity and more exuberance. One is like a cathedral and the other like a theatre. The shaggy boles rise to a height of thirty or forty feet, and the marshy ground between, into which a man may slip thigh deep, bears a strong growth of aroids, *Heliconia* and other swamp-loving herbaceous plants. As Rodway (39) points out, palms are among the conquerors in the battle of vegetation in the rain-forest. They brook no undergrowth which is strong enough to compete with them in the upper levels where their radiating crowns of ten-foot fronds are spread to the light and air. Lianas and vines are almost lacking, for the straight trunks and glossy pendant leaves offer no hold to them, nor to the strangling parasitic climbers which often throttle the great foliage trees. Nothing but a few epiphytic ferns can obtain a footing on the palms.

Mangrove swamp is in some respects so peculiar that it deserves a place to itself, but for the sake of convenience it is included here as a division of inundated forest. Mangroves are plants of coastal regions only, for they are true amphibians and can flourish only where their roots are periodically submerged and laid bare by tidal action. They are inhabitants of the tropics of both Old and New Worlds, but South America is relatively poor in species.

Rhizophora mangle, the form found in Guiana, grows along the river banks in stinking mud at the foot of the high forest. The trees are seldom more than thirty feet high, often less, and their growth is slender and straggling, for the trunk is raised above high-water mark by the arching stilt-like roots. The growth of these roots is most complicated and curious. Each arises as an aerial shoot from the base of the trunk, tipped with a thimble-shaped guard, but as it lengthens it curves over in an arch and enters the mud. Side roots are given off from this primary one, but there is none of the overcrowding and confusion that appears in the swamp jungle. It seems almost as if every tree had a calculating brain, which, acting in co-operation with its fellows, planted each root in its allotted place without interfering with its neighbours.

The mangrove is viviparous. It is clear that in its amphibious station if the seed fell in the usual way it would be swept off and lost in the rising tide. The cotyledon is developed inside the fruit, and a stiff green root, twelve to eighteen inches long, is produced. Later the young plant separates from the cotyledon which remains behind in the husk, and drops plumb like a spear into the mud, where it is rooted by its own weight before the tide can float it away.

Compared with the forest behind it, the mangrove swamp is

a light and airy place. The trade-wind from the sea blows through it, and sunshine can enter. Each tree is a little cosmos in itself, and besides its own pseudo-parasitic fruits, supports a load of orchids, bromelias and other epiphytes. These are visited by bees and other sun-loving insects, and are infested with ants' nests. The mangrove's trellised roots are also visited by birds which never penetrate into the shady forest behind— fly-catchers, tyrant-birds, kingfishers, herons, ibises, the snake-bird, and the sun-bittern with its curious plumage. Lizards and water-snakes also find a safe asylum there. The most noticeable insect absentees are the termites. These ubiquitous and flourishing insects, whose colonies load the trees and cumber the ground in the neighbouring jungle, cannot invade a forest where there is no soil, and where dead timber is swept off by wind and water before it can decay. For those plants and animals which live in the swamp are, with the mangrove tree, slung in the air like Mahommed's coffin. "Earth" to the aerial forms means a maze of roots and branches whence a false step brings death in the abyss of mud or water below. "Water" to the aquatic forms implies a shifting shallow medium which is liable to recede and leave them high and dry. Hence many even of the fish of these tidal shores seem to belong to both elements. Such are the small fresh-water flying fish (*Carnegiella*, *Gasteropelecus*, etc.). At low water the great catfish push grunting on to the mud; and when the tide is up, the curious *Anableps tetrophthalmus* skims the surface, with bulging eyeballs thrust into the air[1].

Conditions somewhat analogous to those in the mangrove swamps are found along many South American rivers where low trees spread out over the water and support a considerable population in their matted branches. Thus a parallel is afforded by the leguminous thorn tree *Drepanocarpus lunatus* which forms thickets along certain streams in Guiana, and is the haunt of that interesting bird, the hoatzin.

The peculiar structure of the hoatzin possesses characters which in some instances may be primitive, but in others are

---

[1] The upper part of the eye of this fish is adapted for aerial, and the lower for aquatic vision; hence the popular name of "Four-Eye."

certainly adaptive[1]. The distribution of feathered tracts in the skin (pterylosis) is less well defined than in most modern birds, and may be derived from an early condition of continuous feathering; but the thickened skin along the breast is probably correlated with the bird's habit of squatting close to its perch. The breastbone is of remarkable form with the keel cut away anteriorly and the merrythought (furcula) completely fused to the coracoids on one hand and to the sternum on the other. It is provided with a median splint of bone which Parker regarded as the homologue of the inter-clavicle of reptiles. The arrangement of the intestines has some resemblance to that of the ostriches and screamers, both considered to be archaic types, but the enormous muscular crop and reduced gizzard are probably adaptations to the bird's diet of leaves.

The above remarks are somewhat of a digression, but are perhaps justified in that a good deal has been written of the primitive position of the hoatzin. As the habits and œcology have been called in support of this view, it is well to point out that in structure at least the bird has evidently undergone a certain amount of environmental adaptation.

The adult hoatzin somewhat resembles a small brown and chestnut pheasant with a long wispy crest. Its plumage is slatternly, its carriage ungainly and its voice a gutteral croak. Other observers have commented on its unpleasant smell—one of the local names is Stinking Anna or Hanna—but I was not able to detect this. The birds, which are most active in the early morning and late afternoon, are gregarious and rather quarrelsome. Their flight is clumsy and feeble, and they seldom, if ever, visit the ground or leave their native thickets. They feed chiefly on the foliage of *Drepanocarpus* and of a giant aroid, *Monotrichardia arborescens*, which grows in the same places; and their life is spent entirely among the slender boughs whose thorns afford them protection against all enemies.

The contrast in behaviour between the adult and nestling hoatzin, of which Beebe (6) has given a full account, is very striking. The young bird, hatched almost naked, finds himself

[1] The following account of the structure of the hoatzin is taken from Beddard (5).

B

4

from the first in a dense thicket, slung over the water. His only means of progression is by climbing, and to this end the unfledged wing has become a grasping organ of some dexterity. The thumb and index finger, each provided with a claw, are capable of a certain amount of lateral movement, and grip the twigs in such a way that the bird, assisted by the feet and using the head as a lever, can readily travel through the branches, though on the ground he is almost helpless. But the adaptation goes further than this, for when alarmed the little hoatzin dives straight into the water many feet below, and, swimming with ease, remains concealed until the danger is passed. This power to climb and instinct to dive, developed in the nestling alone, is probably correlated with the peculiar environment of the early life, and is not necessarily an ancestral character.

The term "higher lying forest" is used here for all the forest area which is not periodically flooded. Any account of this formation is necessarily inadequate, for apart from its enormous extent, it contains many secondary formations which elude definition at present; and explorers, ancient and modern, have already exhausted their vocabularies in attempts to describe it.

The trees are as large as those of the swamp jungle, but it is difficult to estimate their girth because the intermediate space is filled with a host of other plants. There are palms, aspidistras, straggling undershrubs, anæmic saplings, lianas of every size from that of a bootlace to that of a man's thigh unable to stand alone but dragged up to the sun by the trees, aerial roots dropping down as far as the lianas climb up to suck sustenance for the parent plant overhead. The ground is cumbered with rotten stumps, broken branches and fallen ants' nests. Sometimes the "bush-ropes" or "monkey cables," as the woody lianas are called in Guiana, break and fall to the ground to add to the general confusion. It is possible that the forest varies in penetrability in different places. Thus André[1] denies the assertion of many travellers that it is impossible to force a way without an axe, and says that on the Orinoco River the settlers have a special name for patches of dense jungle, and avoid them. But in any case the number of other plants that even a small tree can support is overwhelming. One eight-foot sapling, chosen at random, bore a burden of eleven different kinds of ferns, vines and bromelias, beside lowlier growths innumerable, and the large trees each hold aloft a veritable hanging garden of Babylon. Besides climbers like *Philodendron*, *Norantia* and *Sarcinanthus*, which cling to the trunks like ivy, every notch in the bark and every branch bears a load of ferns, orchids, bromelias and tillandsias, epiphytic or parasitic. Some, like the *Monstera*, which has huge semi-palmate perforated leaves, change their habit in the course of growth. *Monstera*, beginning its life with roots, later becomes an epiphyte, and in course of time thrusts down a new root system. Others are deadly enemies to the trees. A species of *Loranthus*, allied to the mistletoe, has viscous seeds which are eaten by birds and disgorged elsewhere among the branches. The seed germinates, and, instead of a root, produces a sucking disc which draws nourishment from the tree. As the parasite grows, it develops more and more suckers which spread insidiously among the boughs. So successful is the *Loranthus* that when once established it can scarcely be eradicated unless the whole infected area of forest is felled. The fig, *Clusia insignis*, destroys its victim as surely but more crudely than the *Loranthus*. The fruit is dropped by a bird into the fork or cleft of a large tree, and as it germinates and grows, it thrusts down roots to the soil. The host is gradually enveloped and slowly strangled until the *Clusia* stands triumphantly alone, supported by its own intricate root system and holding in its thousand-armed embrace the rotting corpse of the forest giant.

The ground level of shade forest is almost flowerless. Occasionally a scarlet passion-flower or yellow catchfly vine hangs over the trail, and a pretty little creeper, *Geophila* sp., bearing twin berries of the same size and colour as violet flowers, may overrun the ground. There is also a low shrubby plant, *Cephælis violacea*, common in the Guiana forest, whose small white blossoms would be overlooked if it were not that they are borne at the apex of a compact bulbous sheath of bracts, as big as a plum and of a fine purple colour.

So much has been written of the exuberant life of the tropics that a newcomer is apt to be disappointed in his first expedition into the bush. The vegetation may well surpass his expectations in the luxuriance and magnitude of its growth; but he may or may not find at once those huge or bizarre animal forms which travellers' tales have taught him to expect, and frequently the general population seems to consist of types which do not differ widely from those of Europe. The fact is that in anticipation of the great, we are apt to overlook the small; and the great *Papilio* and *Morpho* butterflies when they pass excite such wonder in the beholder that for each of them his eye disregards twenty little Erycinidæ or Pieridæ. But as his experience grows, he will become increasingly conscious of the vast extent of the forest population. Over and above the rustle of foliage, it is possible to detect a faint murmur of movement on a gigantic scale. A million beings contribute to this stir—termites scrambling down their galleries, leaf-cutting ants in procession down the trails, crickets stridulating under the leaves and shield-bugs probing for sap. At times, when the sun shines, cicadas sing like steam saws and wasps clang to and fro in loud chorus, but the undercurrent of small things is still discernible. This bustle of insect life has a curious effect. It is independent of the presence or withdrawal of man; and to an overpowering degree it impresses him with its aloofness and mystery—as of a life whose events and trend are always out of sight and out of reach. In fact the outstanding impression of the tropical forest is of the contrast between the plant and animal life. The first encompasses the beholder at every turn and oppresses him with its exuberance. The other though so near yet seems so remote and so elusive. The psychological effect of this sense of human isolation is curious, and akin perhaps to the loneliness which some temperaments suffer in great cities.

In the rain-forest clearings may arise naturally, as when a tree falls and shatters a space in the surrounding jungle, or they may be due to human agency. The aboriginal and half-breed population of Guiana live largely on meal prepared from the root of the cassava plant. This crop can be raised only for a year or two in the same spot, so that old clearings are constantly abandoned and new ones made.

When the tall trees are felled, the immediate effect is that of a city which has been devastated by fire or hurricane. The inhabitants linger for a while round their homes, bewildered and famine-stricken, and then perish or emigrate. Thus for a day or two the new clearing, cumbered with shattered timber and foliage, contains shade-loving butterflies (Heliconiinæ and Ithomiinæ) and grasshoppers (*Rhipipteryx*, etc.), while arboreal ants and bees, suddenly exposed to the noonday glare, swarm sullenly around. These then disappear and their place is gradually taken by sun-loving butterflies, such as species of *Adelpha* and *Eudamus*, various "skippers" and the tawny and black Nymphalid *Dione vanillæ*. Instead of antbirds and tinamous, anis, kiskadees and grass-finches appear. Where the soil is bare, burrowing wasps (*Monedula* sp.) replace shade-haunting Hymenoptera, and *Rhipipteryx* gives place to green grasshoppers. The area lies derelict only for a short time. In a month or two the broken stumps send up fresh shoots, and grasses, vervains, sages and pinks appear. Where the artificially cleared land is under cultivation, cassava, plantains and pineapples may be introduced. The fauna changes with the flora. Many of its constituents are adaptable species from the surrounding forest which can accommodate themselves to new conditions, but besides these there are a number of forms known only from cleared ground elsewhere. Study of collections from different parts of the forest region shows how constant and remarkable immigration is. Records of captures from localities many miles apart and separated by expanses of virgin forest as impassable as the sea to all but strongly flighted forms, bear witness to the competition for new territory. This is very marked in respect of some of the Homoptera, such as some of the Jassidæ and Membracidæ, and in certain butterflies. The showy red and black *Victorina amathea*, for instance, is not found in the forest in Guiana, though it is widely distributed in the cleared and cultivated coastal lands, and ranges up the rivers wherever trees have been felled and grass has appeared in the neighbourhood of settlements. The conquest is usually gradual, but sometimes a sudden invasion takes place. Beebe (6) describes the arrival of a number of Guiana house-wrens at a clearing where they were previously unknown. "A wave of wrens...in the early morning were feeding and singing and climbing about the grass stems perfectly at home....Within three days they had dispossessed some finches, nest, eggs and all, and had begun nests of their own."

Where the clearing is allowed to revert to forest, the vegetation appears in regular sequence. The first tree to spring up is the *Cecropia* or "Monkey Pump," a slender quick-growing tree with a terminal tuft of palmate leaves. Below, the herbaceous growth becomes taller and stronger, mingled with woody shrubs, bananas, paw-paw trees and pine-apples. Over all scrambles a straggling climbing grass whose stems cut like razors. This thicket grows deeper and even more impenetrable than the jungle around, so that it needs a cutlass to force a passage. As it develops, shade-loving animals are drawn to it in increasing numbers and clearing forms become scarce. The insect fauna of high secondary growth, especially of that which has itself been cleared in places, is rich and varied, more so than that of dark untouched forest. As Forbes (21) wrote in the Malay Archipelago: "The ornithologist and entomologist obtain most of their treasures in the small virgin forest patches in the neighbourhood of villages, in wide shady paths in the great forest, and along sunny walks amid the opened portions of the second growth."

It is not known how long it takes one of these cleared tracts to revert to its original condition. Chapin (18), writing of the Congo forest, hazards a guess of eighty years, and even in a region of maximum heat and moisture the time required cannot be far short of this. It may even be longer.

CHAPTER II

So far these descriptions of the jungle apply only to the forest floor, in the spatial dimensions of length and breadth, but the life of the rain-forest ranges up into the trees, tier above tier, and must be considered in terms of height also.

In temperate woodlands only comparatively few and highly

specialised forms are adapted to flourish above the earth's surface, but in tropical jungle both animal and vegetable life is as abundant and diversified above as below. As Humboldt exclaimed: "Forest is piled upon forest." Each tree is a microcosm, the prop and sustainer of a host of living things which in many cases are not living there fortuitously, but are delicately adjusted to their environment. The tropical forest may be compared to the sea. The human being, prowling about among the roots, is the bottom-crawling benthos. Above him, arranged in regular sequence, are a series of complexes or associations extending right up to the surface of the ocean—that is to say to the tops of the trees. This level is a world in itself, and one of which we unfortunately know as little as the flat-fish on the sea bed knows of the plankton that floats and the nekton that swims above him.

Some butterflies scarcely ever come to earth. From the clearings, great *Morphos* and *Papilios* can be seen sailing like birds over the tops of the trees, but they seldom descend within reach of the collector's net, and there are crickets, beetles, birds and even mammals which live in the same way. A faint idea of the difference between tree-top and ground fauna may be gained by searching the foliage of newly felled trees; but many animals leap off, or are knocked off, in the crash, and even of those which are recovered it is not always certain that they may not have come from the intermediate levels. Bees and stinging ants generally prevent the climbing of trees; hence our knowledge of this "life-zone" is just enough to show us how much we miss.

In French Guiana, wooden scaffolds have been erected for the capture of the butterflies whose iridescent wings are sold in Paris to ornament brooches, pendants and knick-knacks. Mr J. J. Lister tells me that he recently met a merchant of these wares, who was himself a keen lepidopterist and had obtained many valuable specimens in this way. I am not aware that the plan has been tried elsewhere for scientific purposes, but it would probably prove well worth while. If St Simon Stylites had established himself in the Amazonian forest, he might have added considerably to our knowledge of the fauna.

Some of the tree-top forms are beautifully adapted to their way of life. Howes (23) studied the development of a Trypetid fly, *Spilographa* sp., whose larvæ fed, several together, in a tough-skinned red fruit which had fallen from a tall tree. These fruits were much sought after by monkeys and parrots who threw them down in quantities. When found the larvæ had consumed all the pulp within the rind and were full-grown and ready to pupate. When Howes extracted two larvæ and placed them in a vessel full of damp earth, they immediately burrowed underground and transformed there. Two days later he repeated the experiment and once more the larvæ pupated. This happened again and again, while the rest of the brood left in the empty rind remained unchanged. The fruit was collected on April 20th, and on May 12th, when the kernel began to sprout, the last larvæ were placed in the soil and pupated as their predecessors had done. Thus metamorphosis was postponed for over three weeks until conditions were favourable. The advantage of this is obvious. The larva is ready for the change as soon as its fruit habitation is ripe, but it must have access to the soil. As the rind is too tough to pierce, it must wait until some outside agency sets it free. Sometimes a monkey or parrot, attracted by the bright colour, bites the fruit and then the larvæ tumble to earth and burrow at once. If not, they delay until the fruit falls of itself, when it will either burst open in the descent or soon rot away on the ground. In any case the larvæ will ultimately escape, but by their power of marking time as it were, they are able to take advantage of the fortunate accident of monkey or parrot. If their maturity coincided only with the natural fall of the fruit, it might happen that half-grown broods would be precipitated earthwards before their time and perish.

As the flat-fish in the sea is constantly rained upon by the dead and dying denizens of the surface waters, so the naturalist in the forest finds strange objects which have fallen from the tree-tops. Curious bright flowers and fruits drop to earth in this way, and sometimes turn a dark forest trail into a carpet of colour. The blossoms are seldom those of the forest trees, whose flowers, though dependent for the most part on insect pollination, are usually rather insignificant (4, 21, 39). They belong more often to the canopy of vines and creepers which cover the roof of the jungle. Sometimes at night a wave of entrancing perfume floats down from the same upper heights, from a spray of blossoms hidden overhead, which perhaps no botanist has ever named.

Occasionally animals filter down in the same way. A peculiar caterpillar, furnished with long complex spines, used to appear in the vicinity of the Tropical Research Station at Kartabo. One form was black and coral, another was white and blue, but they could never be reared since they were always taken at large on the ground and their food plant was unknown. But one day a large tree was felled, and still clinging to the topmost branches were half a dozen caterpillars of this type, white and brown and gold.

It is no wonder that arboreal animals are so numerous in the rain-forest, for the tree-top level offers an abundant supply of those staple foods, fruit and termites. The neo-tropical forest is the great metropolis of tree-living termites. It has been suggested that the heavy rains and consequent saturation of the soil have driven these social insects upwards out of the way of floods; and this may be a true explanation, for in the dryer savanna forests, ground nests are the rule. Be this as it may, this source of easily obtained food has certainly been a factor to draw higher animals up into the trees. Arboreal habit may be either obligative or facultative. Forms such as the sloth are almost incapable of progression on the ground; while opossums and rats, though nimble climbers, are still terrestrial. Many tree toads are facultative climbers and may be found either on the forest floor or clinging to the leaves of *Monstera* and other creepers forty feet aloft. The accompanying scheme is compiled from the list of mammals of the Bartica district of British Guiana published by Beebe (8). Omitting the manatee, dolphin, and the bats, it will be seen that of fifty-nine species, thirty-one at least belonging to five out of the six families represented in the fauna of the region are arboreal; and of the remaining twenty-eight species, five belonging to four different families are amphibious, a significant fact in forests liable to inundation. It is also remarkable that the prehensile tail, that *ne plus ultra* of arboreal adaptation, is in the main a New World development. All the monkeys on this list, the spider monkey, the capuchin, the red howler, the sackiwinki, the marmoset, etc., possess this third hand, and so do the porcupines, climbing ant-eaters, coati-mundi and kinkajou. Arboreal adaptation extends even to the insect world, for Bates (4) remarked that terrestrial forms of carnivorous beetles (Geodelphaga) are scarce and that most of this group found at Para have the feet modified for climbing.

Table to show the relative numbers of cursorial, arboreal and amphibious mammals of Guiana

| FAMILY | TERRESTRIAL SPECIES | | ARBOREAL SPECIES |
	Cursorial	Amphibious	
Marsupialia (Opossums)		*Chironectes* (1)	5
Rodentia	Agoutis (2) Paca (1) Spiny rats (2) Rats and mice (8)	Capybara (1) Paca (1)	Porcupine (1) Squirrels (2)
Edentata	Armadillos (4) Ant-eater (1)		Sloths (2) Ant-eaters (2)
Carnivora	Hunting dog (1)	Otters (2)	Cats (5) (facultative) Raccoon (1) Coati (1) Kinkajou (1) Tayra (1) Grison (1)
Ungulata	Peccaries (2) Deer (2)	Tapir (1)	
Primates			Monkeys (6) Marmoset (1)

Beebe (6) has a good deal to say of the arrangement of colour in the jungle. He tabulated the species of birds found in one forest area; and then, reckoning the gradations of light from the dimness of the forest floor to the full sunlight of the tree tops as between 1 and 10, he obtained the following result:

| Estimated degree of light | 1 | 3 | 5 | 10 |
| Percentage of bright birds | 0 | 8 | 50 | 83 |

Hence the most brilliant birds live in the fullest light, but this rule holds only for the *colour*, not for the *conspicuousness* of the individual in natural surroundings. Among insects the rule that brightly coloured forms frequent light places does not always apply. Some of the most brilliant of the small bugs live in deep shade. Yellow and red and black and metallic green are quite general among the Membracidæ, Cercopidæ and Jassidæ of dark jungle, while the forms characteristic of clearings are more often obliteratively coloured green or brown. On the other hand, many of the shade-loving butterflies are inconspicuous. Some indeed are mere ghosts of butterflies. The genera *Callitæra*, *Pierella* and *Hetæra* (Satyrides) in some cases have the wings denuded of scales so that they resemble bits of transparent mica. The same thing to a less degree is found in certain Erycinidæ (*Tmetoglene*, *Stalachtis*) and in some Ithomiinæ. Others have a full covering of scales, but are obliteratively coloured. Some species of *Anæa* (Nymphalides) float low over the ground among the undergrowth and feed on the moisture exuding from dung and decaying vegetation. The wings are beautifully glossed with purple on the upper surface, and when folded resemble a dead leaf in form and colour. The same twilight camouflage of purple and grey is seen in some *Eutychia* (Satyrides) which fly low in the same places. The finest of the neotropical family Brassolides are the great "owl-butterflies." They are soberly coloured, and remain at rest by day, but flit at dusk along the forest trails like the bats for which they are easily mistaken.

The shade forest however possesses some conspicuous butterflies of its own, such as the tawny and black Ithomiinæ and Heliconiinæ which sometimes come together in parties round a favourite bush. As Bates, with his usual power of happy description, remarks: "Their elegant shape, showy colours and slow sailing mode of flight, make them very attractive objects, and their numbers are so great that they form quite a feature in the physiognomy of the forest, compensating for the scarcity of flowers."

The upward search for food and the pursuit of enemies have led to the overcrowding of the "attics" of the jungle. At least so it appears from the number of nests, cocoons and so forth which dangle by a thread in mid-air. Bates(4) commented on the number of pendant moth cocoons in the Amazons' forest, and pointed out that they are well adapted to withstand the peck of a bird. Belt(10) considered this device as a defence against marauding ants, and remarked that of all small animals, spiders in their webs stand the best chance of evading the hunting columns of army ants. Many birds, fly-catchers, orioles and honey-creepers, build pendant nests; and so do certain solitary wasps, which construct exquisite little purse-shaped nests of vegetable fibre, slung by a flexible stalk and stocked with Collembola. Most remarkable is a Chalcid, an internal parasite of butterfly larvæ, which when about to pupate leaves the empty skin of the host and makes a cocoon the size of a cherry stone which dangles at the end of a thread four or five feet long.

Apart from the rivers and their tributary streams, and the curious case of the reservoir plants which will be described presently, the rain-forest is poorly furnished with permanent standing water. Lakes or ponds of any size are rare, and the small pools tend to fill up rapidly with vegetable debris and are transformed into mere boggy patches. However, in the rainy season, pools are formed wherever the soil is non-porous enough to retain water, and these puddles speedily acquire an aquatic fauna of beetles, water-bugs and dragonfly and mosquito larvæ. All these, it may be remarked, are either winged forms, able to migrate when conditions are unfavourable, or larvæ which will develop into imagines at the onset of the dry season, but there are also certain crustacea and other less mobile animals which are able to travel some distance over damp leaves in a moist atmosphere. Such are certain small fish of the genus *Rivulus*, which are sometimes found in rain pools at considerable distances from permanent water. The forest frogs and toads take advantage of the rainy season to lay their eggs, but the danger of desiccation is ever present and there are many adaptations to guard against it. Some species carry their eggs or tadpoles from one wet place to another. The best-known example of this is *Pipa*, the Surinam toad. The female carries the eggs on her back in little pits in the skin, each of which is provided with a fold of the epidermis for a lid; but specimens in this condition are said to be rare in collections, as at the beginning of the rainy season the toads withdraw into the recesses of the forest. *Pipa* is a flattened toad with a somewhat pointed head and black warty skin, and the fingers are furnished with little star-shaped discs. It is a water-loving animal, and the only example I found in Guiana was dredged up with the dead leaves from the bottom of a small pool. In the hand it was inert and sluggish, but was capable of a lightning leap when alarmed.

A pretty little scarlet and black *Dendrobates*—a ground-living species—carries the tadpoles on its back. In this case the male is the nurse, as in the European midwife toad; but in *Hyla evansi* the female transports the eggs on her back, and also the tadpoles, which are said to have a primitive type of air-breathing gill. Other toads provide differently for their young. *Leptodactylus mystacinus* lays her eggs in a heap of froth beside a temporary pool, and the tadpoles when hatched wriggle about in the mass. Thus, even if the pool dries up, they are protected from enemies and assured of moisture until their lungs become functional. *Phyllomedusa bicolor* lays her eggs in a leaf just before the rains in such a way that the tadpoles will be washed out by the downpour to develop in a temporary pond.

The most remarkable water system in the forest is that of the reservoir plants. Some forms, such as the banana and *Heliconia*, carry a little water in the axils of their leaves or bracts, but the most interesting reservoirs are those of the Bromeliaceæ. The common pine-apple is the best known member of this family, but it is a ground-living form. The greater number of species are epiphytes which grow on the forest trees at any height from one foot from the ground to one hundred. The leathery leaves are arranged in a stiff rosette round a cavity, which even in the dryest weather contains a quantity of water. These bromelia basins are often the only water available for a host of aquatic and semi-aquatic animals. In short, to quote the picturesque phrase of Picado, the whole bromelia flora of the forest is a "grand marécage fractionnée," a great discontinuous marsh, whose units are separated in space and raised above ground level. The trees bear thousands upon thousands of little aquaria, each holding half a pint or so of water, and these plant cups take the place of the larger, less numerous pools of terrestrial marshland.

The bromelia fauna includes facultative forms which wander in for food and shelter, and obligative forms which can live nowhere else and sometimes are specially adapted to their peculiar environment. Representatives of the following groups have been recorded:

> Protozoa.
> Rotifera.
> Turbellaria.
> Hirudinea (leeches).
> Oligochæta.
> Gasteropoda (snails).
> Crustacea (Isopoda, Copepoda, Ostracoda).
> Onychophora (*Peripetus*).
> Myriapoda (centipedes, millepedes).
> Scorpionidea (scorpions).
> Arachnida (spiders).
> Insecta.
> Amphibia (frogs, which spawn there).

The list of insects is a long one and includes:

> Thysanura.
> Orthoptera.
> Thysanoptera.
> Hemiptera.
> Hymenoptera (ants).
> Coleoptera.
> Aquatic larvæ of Perlidæ, Odonata and Trichoptera.
> Larvæ of Lepidoptera and of seven families of Diptera.

An even better idea of the variety of this strange aquatic life of the tree-tops can be gained from an actual census of a bromelia chosen at random.

In Costa Rica, Calvert[17] obtained the following forms from a single plant: larvae of dragonfly, a scorpion, two species of phalangids, a pseudo-scorpion, eleven species of beetles and their larvæ, belonging to nine different families, a Stratiomyid larva, two species of bugs, a caterpillar, an earwig, numerous ants, and an earthworm.

In a clump of bromelias in Trinidad, Scott[42] found: two frogs, a millepede, some Isopods, some dragonfly larvæ, an earwig, some cockroaches, three species of beetles, a Thysanopteron and several bugs.

Some of these bromelicolous forms are specially adapted to their way of life. Certain cockroaches and bugs are flattened dorso-ventrally to enable them to creep between the overlapping leaves. Picado[32] found a *Syrrphus* fly larva with ventral suckers by which it could cling to the leaves of its aquarium instead of falling out if the plant was upset. Calvert[16] has shown that certain dragonflies of the sub-family Mecistogasterinæ are reared in bromelias. The body of the nymph is not much elongated, but at the final moult a great extension takes place; and it is suggested that this is to enable the female to introduce her eggs into the water, lying deep down in the leaves.

The peculiarities of the bromelia environment do not end here. It has long been known that the plant has scales at the base of the leaves, by which it absorbs salts and water from its reservoir; but Picado[31] observed that the water was always fresh and clear, and that the organic matter it held did not putrefy. He experimented further and showed that the plant secretes two ferments, an amylase which turns starch into glucose, and a trypsin which converts albuminoids into peptones and amino-acids. These products are absorbed by the plant, which thus lives by a kind of external digestion. Presumably the inhabitants of bromelia water must be adapted to withstand the plant's secretions.

The means of dispersal of life in the forest are little known. The gradual penetration of clearing forms into new territory has already been described, but it is uncertain how they travel. The tropical jungle is by no means a homogeneous region, and some species extend over enormous areas. To take one family as an example, the majority of the Membracidæ of Guiana belong to genera, and, in many cases, to species, which are recorded from Central America on one hand and from the Amazons or even from Rio de Janeiro on the other. Wind cannot be a constant factor in dispersal. Floods probably play some part along the rivers. In heavy spate the current tears up masses of soil and vegetation and sometimes carries them for great distances. Floating islands, acres in extent, have been sighted off the Amazons, and occasionally carry even jaguars and large snakes as passengers. Even when the water level is normal, patches of the aquatic grass *Panicum elephantipes* drift up and down the tidal estuaries of some of the rivers of northern South America. Beebe (6) has seen hoatzins borne along on such grass rafts, and has suggested that the curious local distribution of the bird may be due partly to this agency. The wearing away of the banks would be even greater if it were not for the aroid *Monotrichardia arborescens*—the mocca-mocca of the creoles— whose thick stalks fence the shore with a ten-foot palisade, and the mud-binding roots of the courida (39).

Certain butterflies periodically migrate in large numbers. These flights are said to be composed chiefly of males of the genera *Catopsila, Callidryas, Aphrissa*, etc. They belong to the same family as our brimstone butterfly, which they resemble in colour, ranging from orange to cream. Sometimes such a flight passes a given point for hours together. The direction varies according to the locality, for migrating butterflies have been observed on many points on the Atlantic coasts of South America. Belt (10) noticed that in Nicaragua they always travelled south-east, and I have seen a flight towards this point up the Mazzaruni river; but according to Beebe (9) the direction in Guiana is usually north-north-west to the sea. Bates observed a great flock of butterflies crossing the Amazons from north to south (4). Occasionally great numbers alight on the sandy beaches of the river, and make a very beautiful spectacle with their yellow petal-like wings massed together like a bed of bright flowers. I once walked unexpectedly into a cluster of these butterflies. As they rose swirling and eddying in every direction, the sight recalled an autumn scene in Europe when a gust of wind sets the golden leaves dancing through the forest. The meaning of this flocking and the destination and purpose of the migration are not known.

It would be a matter of great interest to ascertain whether the roof of the forest is populated by the same forms as the clearings, and how far the tree-top environment, a hundred feet aloft, is comparable to that of young secondary growth close to the ground. This idea presented itself at the capture, immediately after the felling of a large jungle tree, of a species of frog-hopper which is characteristic of the low clearing vegetation of the region. In this instance it was not certain that the specimen came from above; but if in future it is found that such forms can exist in the upper foliage, it would go far towards solving the problem of their dispersal.

In some respects the roof of the forest may be compared not only with the clearings, but with a prairie or savanna. There is the same wide green expanse, strewn with flowers and open to sun and rain and wind. Butterflies hover and grasshoppers skip over the surface, and its denizens are exposed to the unrestricted view of birds of prey—vultures, kites and harpy-eagles—which soar over the forest, ready to seize any bird or monkey which is not alert enough to dive under the foliage. Parrots, parrakeets, toucans, caciques, and cotingas among birds, and monkeys among mammals, belong in particular to this life-zone, and are generally invisible from the ground. The awe-inspiring boom of howling monkeys may be heard twenty times ere the troop is seen. The little niggling chatter of parrots round a fruit tree can be heard all day long; but the birds themselves are out of sight unless perhaps at evening when they fly across the river to roost, travelling in couples, mate with mate. Toucans are eminently arboreal, and are seldom seen in the lower levels. Bates relates that he once shot one of these birds, and at its cry he was immediately surrounded by a flock of toucans which appeared as if by magic in the thicket where previously not one had been seen.

If the parallel between grasslands on the one hand, and tree-tops and clearings on the other, is to hold good, a proportionately large number of social species might be expected to occur

B

there, for open country as a rule is the home of gregarious birds and beasts. However, this does not seem to be the case. True, monkeys and the birds just mentioned, together with the anis, tanagers and seed-finches of the clearings are social, but the jungle floor has its gregarious species also. For instance, the white-lipped peccary travels in herds of a hundred strong, and when attacked is said to be one of the most dangerous animals in the forest. Birds of the deep jungle unite in flocks which are often composed of several species. These mixed parties are very characteristic, and have been commented upon by several travellers. The forest at first seems quite empty of bird life, and then suddenly the undergrowth becomes alive with Dendrocolaptidæ, barbets, ant-birds, manakins, etc. These all work in the same direction, and presently pass out of hearing, leaving the place deserted again. It is doubtful whether abundance of food or the need for mutual protection is the tie that keeps these mixed companies together. A tree with ripe fruit is exploited by all the frugivorous birds of the neighbourhood; and insectivorous species may be attracted in the same way by local abundance of favourite food, such as a swarm of termites, or the insect refugees from a raid of army ants.

Any description of the rain-forest is inadequate that contains no mention of its marvellous bird calls. These for the most part belong to the zone of the jungle floor, or to that dim and little-known region that lies between the observer's head and the foliage roof. Here, where the tree trunks spring up stark and straight, as if to shake off the lianas that threaten to strangle them, and the gloom is scarcely less than it is at ground level, is according to Beebe (6) the metropolis of bird life. He enumerates thirteen groups whose headquarters lie between twenty and seventy feet aloft—curassows, pigeons, jacamars, trogons, honey-creepers, etc. Some of these have tremendous voices. The bell-bird tolls a single sonorous note. The calf-bird lows like a cow. The trumpeter has a call like a cracked trombone, and a big ground dove makes a sound like the gurgle of water, which is audible at a great distance. At dusk the nightjars come out, and the forest falls asleep to the ghostly reiterated "Who-are-you" of *Nyctidromus albicollis* and the goblin three-syllabled

PLATE II

PALM SWAMP

cry of *Nyctibius griseus*. The low undergrowth has its vocalists, and its instrumentalists also. The most curious of the latter are the crackling manakins. When alarmed they beat their wings with a sharp switching noise which sounds like the crash of a large animal breaking through the jungle and is sufficiently alarming to the uninitiated listener. But the sweetest songster of the Guiana forest, and perhaps in the world, is the quadrille wren which haunts low dark undergrowth. No pen can do justice to its delicious melody, which is neither a whistle nor a warble but has a mysterious finished charm unlike that of any other bird music with which I am acquainted. Clear and deliberate, it suggests some elfin singer, sitting aloof in the dim thicket and improvising on silver panpipes.

The tinamous belong to the ground level, and their calls, clear and far reaching with some of the passionate yearning of the curlew's spring song, are among the most characteristic sounds of the forest[1]. The tinamous are restricted to South America, and superficially resemble partridges in plumage and behaviour, but structurally they are distinct and probably of an ancient type. The three species round the Essequibo river have been studied by Beebe (6), and two of them have curious nesting habits which recall those of some of the ostrich tribe. Each lays but a single egg at a time, and this is incubated by the male, who also rears the chick. No sooner is the latter able to fend for itself than this devoted parent calls for another mate to provide him with another nursery. Indeed in some cases he has been known to incubate an egg while still attended by the three-parts grown chick of the previous brood. The great tinamou is more normal in its nesting behaviour, but differs from the other two species in roosting in trees, a curious departure for a bird of essentially terrestrial habits. In correlation with this, the posterior surface of the tarsus is roughened, while in the two smaller species which roost on the ground it is smooth. These differences in habit and structure were first pointed out by Waterton (47) and have been recently confirmed by Beebe.

[1] This resemblance constantly occurred to me while listening to the trilling of the tinamous in the Guiana forests, and W. H. Hudson, as I found later, had already remarked the same thing of the tinamous of the pampas (*Adventures among Birds*, London, 1913).

A feature of the rain-forest is the number of species present. In arctic or sub-arctic regions it is usual to find a great number of individuals occupying the same area, but as a rule these belong to comparatively few and highly specialised forms. In Brazil a census of three square miles of forest showed that the trees belonged to no less than four hundred different species. The British Isles is an area of considerable œcological variety, and the avifauna is comparatively rich; but ten miles of Guiana forest and river bank have yielded four hundred and sixty-four birds—considerably more than the whole number on the British list (8).

It may be remarked here that a first impression of the tropical forest sometimes suggests that although there are a great number of forms, there are on the whole fewer individuals of a species. However, this is probably human error, and due to the complex and crowded environment which makes anything like a complete survey very difficult. If the observer settles down to study one group intensively, he will discover such a wealth of life as will soon convince him of his error. It is this complexity of environment which makes the jungle the worst place in the world for the study of œcology, since besides topographical and climatic influences, every animal is involved in intricate relations with a host of others.

Questions that are sometimes asked are: Is the struggle for existence in the tropics fiercer than elsewhere? and does evolution—the production of new forms—take place at a quicker rate? The answer to the first question is, strictly speaking, no. Perhaps an animal is liable to attacks by more enemies than elsewhere, but set against this, famine, drought and cold are unknown. It is true that there are more mouths to fill with the abundance of good things, but this does not really affect the matter. Every region supports a mean population whose size is determined by the supporting capacity of the land at the least favourable season of the year. In the rain-forest the seasonal changes are slight compared with those elsewhere, and there are always warmth, light, shade and moisture.

To the second question the answer might be yes. The environment is more complex owing to the greater number of forms of life it contains, and it is justifiable to suppose that more adaptations have arisen to meet it. Moreover, though the rate of reproduction is not proportionately higher than elsewhere, it is more evenly distributed in time. Breeding is not necessarily confined to a single season of the year. Many birds, perhaps the majority, breed twice in the twelve months; and although they lay as a rule half the number of eggs to a clutch of their allies in the north, this possibly gives more scope for crossing to occur. The death-rate is high, and if the survivor of a couple has to mate again for the next brood of the year, this will give a second chance for a re-shuffle of the factors which give rise to variation (or mutation).

Let us suppose two hypothetical animals, A from the arctic and T from the tropics, which, to maintain the racial mean, must each produce six offspring per annum.

A in the short northern breeding season takes a mate, A_2, and they rear six young. Hence: $A + A_2 = 6$.

T also takes a mate, T_2, but as under tropical conditions they breed twice a year, their first brood is of three only. For the second brood there are two possibilities. Either T may pair with T_2 and rear three more young, which case may be expressed: $T + T_2 = 3 \times 2 = 6$. Or T_2 may be killed before the second brood is produced and T must take another mate. In that case his annual contribution to the population might be expressed:

$$(T \times T_2 = 3) + (T \times t_2 = 3) = 6.$$

Thus, though the number of offspring is the same, the descendants of T have a more mixed parentage than those of A, and the possibility of re-arrangement in the constitution of the next generation is doubled.

A point in the death-rate of tropical animals, which as far as I know has never been raised, is their resistance to the action of bacteria on injured tissues. In hot moist climates it is a commonplace of human hygiene that septic conditions are readily set up, even in trifling abrasions, and that neglect of a wound which could safely be ignored elsewhere may be fatal. I was led to this consideration by an anecdote of André[1], who shot a tapir in miserably wasted condition. The animal had been clawed by a jaguar some time previously, and its flank had festered and become fly-blown. Sportsmen and others in this country who have had experience of wild animals, know that they often recover even from extensive injuries, and it would be interesting to learn whether this holds for the tropics also. Is "scotched" synonymous with "killed" in that climate, or has the resistance of the organism increased with the risk of infection?

Mr Beebe, who has had much experience of neo-tropical snakes, tells me that these animals require delicate handling, for a comparatively slight bruise or abrasion sustained during capture sometimes leads to death a week or two later from some kind of bacterial infection. This susceptibility is perhaps peculiar to snakes and may not affect other vertebrates; but the need to avoid even slight injuries may account for the readiness for flight and the threatening behaviour characteristic of even non-poisonous reptiles.

Caterpillars offer an interesting analogy. Poulton (33) remarks that the larvæ of Lepidoptera and Hymenoptera possess a great variety of defences, nauseous glandular secretions, unpleasant taste or smell, irritating hairs, and so forth, and that these are frequently advertised by conspicuous colours or a terrifying attitude. He points out that larvæ are for structural reasons particularly susceptible to injury. A caterpillar may be described as "a soft-walled cylindrical tube which owes its firmness and indeed the maintenance of its shape to the fact that it contains fluid under pressure. The pressure is exerted by the muscular parietes of the body...." "This construction is extremely dangerous, for a slight wound entails great loss of blood, while a moderate injury must prove fatal. The larvæ of *Smerinthus ocellatus* (and many others) nibble off each others' horns, and the wounded larvæ, though they do not seem to be aware of the injury, lose a great deal of blood, and although they may recover are generally stunted; and often I am sure the loss of blood proves fatal."

In the present state of our knowledge it would be unwise to conclude that risk from wounds has played a direct part in the evolution of tropical forms as compared with those elsewhere;

but we are justified in the conclusion that in the tropics, defence against rapacious enemies has reached its highest development. The majority of these adaptations, bewildering in their variety, fall into two groups: associations for mutual protection and welfare, and devices for the deception of enemies.

The two following chapters will be devoted to these aspects of the struggle for existence in the jungle.

CHAPTER III

ASSOCIATIONS for mutual protection and welfare may be formed between animals and plants, or between members of the same or of two or more species of animals.

One of the most curious plant-animal associations is that which has arisen between the sloths and certain green algæ of the *Protococcus* group.

The sloths, which are peculiar to tropical America, are sedentary animals, so adapted to arboreal life that they rarely visit the ground[1]. The tail is rudimentary, the limbs are long and powerful, and the senses are apparently poorly developed. The internal organisation is so aberrant that Buffon remarked that with a few more defects of structure a sloth would be unable to live at all. Most of the life is spent in a semi-dormant condition high up in the trees. Here in the gloom of the rain-forest, clinging to the crutch of a branch shaggy with moss, the sloth either escapes notice altogether, or is mistaken for an ant's nest. The pelage is coarse and grey, with an outer coat of long brittle hairs. The hair shafts of the two-toed sloth are longitudinally ribbed: those of the three-toed species are smooth and furnished with elongated scales (44). The algæ grow in the crevices of the ribs and scales respectively, forming patches along the length of the hair. Their abundance depends on the moisture in the atmosphere; and when present in numbers, they tinge

[1] It is however a curious fact that the three-toed sloth at any rate is a fair swimmer, and is not infrequently captured while crossing creeks and rivers.

the whole animal a dull grey-green, which in the dim surroundings is remarkably protective. The young sloth, which is carried about clinging to its mother's breast, is infected by the alga which is thus transmitted from one generation to the next.

The two species of sloth, though they resemble each other closely in distribution, appearance and habits, differ in their diet. Both are vegetable feeders, but while the two-toed sloth has a comparatively wide range of food plants, and when caged readily learns to eat bananas, lettuces, etc., the three-toed sloth feeds almost exclusively on the foliage of the *Cecropia* or imbauba tree, and can rarely be kept long in captivity. This sharp physiological distinction between two closely allied forms may give the clue to some otherwise inexplicable extinctions and survivals in the history of animal evolution. If the *Cecropia* was attacked by disease and disappeared, the probabilities are that the three-toed sloth would vanish likewise, while the two-toed species would survive. Millenniums hence, naturalists would be puzzled to account for the sudden extinction of one form and the survival of the other, for structurally the three-toed sloth is perhaps less specialised, and there is nothing in the anatomy to suggest the gastronomical difference between the two species.

The *Cecropia* is not only the staff of life to the three-toed sloth, but it also gives protection and shelter to various species of ants. *Cecropia adenopus* of Brazil in particular was first studied in this respect. It is inhabited by stinging ants of the genus *Azteca*, which are obligative inhabitants of the *Cecropia* tree. The shoot or sapling is hollow, and above the insertion of each leaf is a groove in the stem wall of the internode. This groove arises mechanically through pressure by the developing bud[1], but if a section is taken across it, it is found that the stem wall is actually altered in structure. Vascular bundles and lignified tissue are wanting in the groove, and the stem wall is reduced to little more than a thin diaphragm. It is always at this point that the foundress queen ant bores her way into the *Cecropia*. The door is in fact provided for her: she has only to open it.

[1] A similar groove may be seen above the node in a bamboo stem.

But the adaptation to the ant association goes further than this. The base of the *Cecropia* leaf is padded with a mat of thick brown hairs (trichilium); and growing up through the latter, and ultimately lying free upon its surface, are numerous white oval bodies, called "Müller's bodies" after their discoverer. They are modified leaf-glands, and being rich in albuminoids and oil, are much sought after by the ants. That they really are an inducement held out by the tree to the insects is suggested by another species of *Cecropia* which is never occupied by *Azteca*, and has no Müller's bodies. The longitudinal groove is present, and its wall, though thin, is identical in structure with the rest of the stem wall. Further, the stem of this *Cecropia* is covered with a waxy substance repugnant to all ants, and this gives a possible clue to the meaning of other plant-ant associations of the neotropical forest. In South America the leaf-cutting or "parasol" ants, of which more presently, are great enemies of vegetation, and sometimes defoliate whole plantations. Schimper and Müller (41) found that *Cecropia* which did not harbour stinging ants was attacked by the leaf-cutters: hence it is worth while for the *Cecropia* to offer food and shelter in return for defence. The *Acacia conigera* described by Belt (10) has hollow stipular thorns in which ants live, and the Müller's bodies are represented by analogous "Belt's bodies" on the leaves. The *Cordia nodosa* of South America is a stout hairy plant in which the upper part of the flowering shoot forms a tuft. The stem immediately below the terminal leaves has a hollow bulbous swelling occupied by colonies of small biting ants, which go in and out by an opening between the petioles.

The association of ants with the *Tachigalia* in the Guiana jungle has been studied by Wheeler (50). *Tachigalia paniculata* is a slender leguminous shrub or tree, seldom growing more than thirty or forty feet high. The petioles are swollen and hollow at their proximal end, and the distal half bears six or eight pairs of drooping lanceolate leaflets. The hollow petiole contains a kind of dry red pith which is constantly being formed by four longitudinal strands of yellow parenchymatous tissue, arranged in two pairs along the upper wall. It is this nutritious amber-coloured parenchyma which is so attractive to many of the *Tachigalia's* insect associates.

The petioles are a refuge for numerous small invertebrates, many of which—spiders, mites, millepedes and beetles—are merely casual guests. This is also true of some ants, for Wheeler found no less than thirteen species which can also live elsewhere. The obligative ant-guests of *Tachigalia* are two species of *Azteca*, and two of *Pseudomyrma*. The fecundated queen bores into the petiole and closes up the hole behind her. Inside, she produces a brood of workers; and the latter re-open the entrance and gradually take possession of the whole tree. In the earlier stages several queens of the same or different species may establish themselves in the petioles, but later on one colony appears to drive out the rest, or, in the case of the same species, perhaps the different broods amalgamate. Soon after the petioles are opened by the young workers, Coccids[1] enter, whether by hazard or by the agency of the ants is not known, and attach themselves to the nutritive parenchyma. As they suck, they secrete rectal "honey-dew" as do the aphides of this country, and this supplies the ants with food. As the plant grows, petiole after petiole is occupied, and where the cavity becomes too large it is partitioned into chambers and galleries. Gradually the whole tree becomes a great ant fortress whose denizens swarm to the attack if the foliage is disturbed. The advantage to the *Tachigalia* is considerable, for Wheeler found that ant-infested examples were never defoliated by leaf-cutters, whereas those whose colonies had died out were sometimes stripped.

Ants are not the only insects who have realised the protection afforded by *Tachigalia* petioles, and the luxury of Coccid honey-dew. Wheeler found that certain petioles on plants whose ant colonies had died out were occupied by a social Silvanid beetle[2]. The male and the female beetle enter the petiole, either together or within a short time of one another. They begin by shovelling the debris of any former occupation up to one end, using their flat heads for the purpose, and then start to feed on the nutritive parenchyma. The Coccids are already in possession, or else come in shortly afterwards. The beetles are thus able to supple-

---

[1] *Pseudococcus bromeliæ.*   [2] *Coccidotrophus socialis.*

ment their diet of vegetable tissue with drops of honey-dew, and the parents and larvæ struggle together to obtain the fluid. Wheeler points out that the antennæ and labium respectively are admirably adapted to caress the Coccid into yielding up its secretion, and to scoop up the globule when it appears. The larvæ grow up and lay eggs in their turn, so that after prolonged occupation the same petiole may contain several generations of beetles in all stages of development. As the old ones die off, the survivors shovel their remains on to the rubbish heap at the end of the cavity. When in spite of their sanitary exertions the chamber becomes overcrowded, the beetles leave in search of a new petiole.

The Coccids themselves are the gainers by this beetle-ant association, for, like other sedentary bugs, they are liable to parasitisation by Hymenoptera, and as their ant partners jealously drive other insects from the nest, the Coccids can live in comparative safety. They do not fare so well in association with the beetle, which is a less careful guardian of the petiole and allows the ingress of a small Chalcid wasp which oviposits in the defenceless bodies of the Coccids and ultimately destroys them.

Another plant-ant association is that described by Wheeler (49) as an "ant-garden." A large brown ant[1], and a small black one[2], make a joint nest of earth on the branch of a tree. The ants live on friendly terms, forage together, and use each other's passage ways, but their broods are reared in separate chambers. When the nest is disturbed, the little black ants swarm forth to attack, and if they cannot put the intruder to flight, the larger species, which is more formidable, joins in. Meanwhile the ball of earth gives roothold to a number of epiphytes, which in return for the protection from leaf-eating animals, afford moisture and shelter to the nest. An analogous case is found in certain orchids, *Diacrium* and *Coryanthus*, whose roots form a hollow bulb or meshwork which is always occupied by ants. When these orchids are taken for cultivation, they are frequently immersed in water to get rid of the ants; and it is said that they never flourish afterwards, as they are particularly liable to be

---

[1] *Camponotus femoratus.*   [2] *Crematogaster parabiotica.*

eaten by cockroaches and other pests which in a wild state are driven away by the ants(39). A large number of other plants in the tropics of the Old and New Worlds possess hollow growths in stems, leaf-bases, and tubers which are obligatively or facultatively occupied by ants. Mention should also be made of extrafloral nectaries on leaves or petioles. Belt(10) suggested that the purpose of these structures might be to attract ants, which, once established, would protect the foliage from other insects and leaf-eating mammals.

This view of myrmecophytism has been much debated during the last thirty years; but there is a growing tendency to regard the relation of the ants to their host plant as one of parasitism rather than of symbiosis, and to believe that the older observers overestimated the value of ant protection to the plant. As Von Ihering remarked: "*Cecropia* can live without *Azteca* as easily as a dog can live without fleas." Chodat(19) even holds that the swellings in stems and petioles of such plants as *Acacia* and *Cordia* are not due to ant association at all, but are gall-like growths induced in the first place by phytophagous Chalcids, and subsequently occupied by ants. This theory has been recently criticised by Bequaert(11).

Bailey(2), from the study of African myrmecophytes, is led to the conclusion that the relationship is one of parasitism; but he points out that the investigators who have succeeded in overthrowing the symbiotic theory have failed to provide sufficient explanation of many peculiar structures associated with myrmecophytism, though on histological grounds he considers that some tissue modifications at least are traumatisms induced by the ants. Bailey summarises the matter as follows: Certain plants, for reasons which are at present obscure, tend to form extrafloral nectaries, food-bodies, fistulose branches, etc., and ants often, though not invariably, take possession of these. In the resulting partnership, the advantage is all on the side of the insects. Nevertheless, no matter how they arise, there is no doubt as to the frequency and intimacy of these plant-ant associations, and their importance in jungle bionomics.

Associations where plants are dependent on animals are rare except as regards the pollination of flowers and the dissemination of seeds. Pollination by birds is unusual, but is known in several cases. Humming-birds are the pollinators of the neotropics, and, according to Belt and others, several flowers are dependent on their visits. It is sometimes supposed that they are attracted to red flowers, such as the species of *Combretia* which they often visit, but one of the most interesting humming-bird flowers, *Marcgravia* sp., has inconspicuous green and brown blossoms. The plant is a climber, hanging over the water at the river side. The florets are arranged in a circle, and in the centre of the wreath are five purse-shaped nectaries, each with an aperture at the base. It is supposed that the birds dart their bills into the nectaries, and that in so doing their plumage is dusted with the pollen. In the Old World, a kind of mistletoe, *Loranthus acaciæ*, is fertilised in the same way by one of the sun-birds, a group which resembles the humming-birds of America in appearance and habits.

Associations among animals are often even more complex than those between animals and plants. Every possible combination is found from the loosest commensalism to the highly organised commonwealths developed by single species; and it is impossible here to touch more than the fringe of the subject.

Sometimes one form is drawn to another for protection. Belt (10) has commented on the choice of certain birds to breed in trees such as the "bull's horn" acacia, which is inhabited by stinging ants. He mentions in particular the case of a fly-catcher which breeds in thorn bushes, usually close to the nest of a banded wasp; and he cites Gosse who says that the Jamaican grass-quit often builds in a shrub occupied by wasps.

Food is sometimes a bond. The bugs of the curious family Membracidæ excrete honey-dew. Many of these insects form colonies on the stems of plants, where they are constantly visited by ants who caress them with their antennæ and entice them to give up the sweet liquid. The ants obtain food and the Membracids profit by their presence, for any interference with the bugs is resented bitterly by the visitors. Some ants build little shelters of earth or vegetable fibre over the Membracid colonies, and at first sight this appears like an intelligent attempt at domestication; but closer investigation shows that the same ants build shelters over Coccids also, and even over nectaries on leaves, and that the Membracids prosper as well with no roof over them. As a matter of fact, the Coccids and nectaries are sedentary and are therefore more satisfactory subjects for imprisonment, for a Membracid when startled is capable of a sudden leap through the roof of its newly constructed pen and so away. I have seen a tree, about the size of a big English holly, of which every twig was covered with Coccids which the ants had laboriously immured in hundreds of little mud kraals. If the tree was shaken, the ants swarmed out so furiously that it was easy to realise the protection that their protégés received in this formicarian stronghold.

A curious instance of combined defence, correlated with structural adaptation, is found in the brood of a Chrysomelid beetle[1]. The larvæ are thickly hairy, and the last segment of the body is expanded into a strongly chitinised, shovel-shaped flange. The larvæ feed in a compact mass on the upper surface of *Cecropia* leaves. Their heads are all directed inwards, while the caudal expansions thresh to and fro on the outside of the circle. The chief enemy of these larvæ is a carnivorous Pentatomid bug[2] which loiters on the outskirts of the throng, awaiting the opportunity to impale a larva with his proboscis and drag it from its fellows. As long as the circle of shovels is unbroken, the bug stands little chance, for his stylets cannot penetrate their polished armour and he cannot reach the soft bodies beyond. But as the larvæ feed, they move outwards from the original centre, the circle becomes wider, and sooner or later the enemy slips in between the defences and secures a victim. According to my observations, however, the circle is not broken naturally until the larvæ are full-grown, in which case the loss of one or two individuals does not signify, for before the bug has digested its meal, the rest of the brood enter the pupal stage and so escape.

The social Hymenoptera—ants, bees, and wasps—and the termites or so-called "white ants," live in polymorphic communities which not only possess a highly developed caste system, but include a great number of other forms—guests,

---

[1] *Cœlomera cayennensis.*   [2] *Phyllochirus.*

commensals, parasitic and predatory enemies, and so forth. The bees and wasps claim no attention here, for on the whole the tropical forms possess no particular characteristics; but in the regions near the equator the ant family reaches its highest development in the size and number of its members, and in the complexity of its organisation. In fact in the neotropical jungle, ants are the outstanding forms of life. Social or solitary, large or small, poisonous or innocuous, under, upon, or above the ground, they swarm everywhere, until the forest seems at times to be a sort of nightmare of ants. If the naturalist brushes the foliage in passage, a score leap upon his garments; to climb a tree is to court a thousand needle pricks or worse; and a pause to rest on the ground is the signal for a general invasion. In camp the irritating insects destroy his specimens and spoil his food. In fact he may conclude that there is more than a germ of truth in Wells's story of the ants who held a whole country against human approach. There is no doubt that ants are a serious factor in the bionomics of the rain-forest, and indeed in the tropics generally. Beebe (6) notes that young birds may be killed in the nest during the temporary absence of their parents. Belt (10) observed that small mammals are exposed to the same danger. He remarked that some of the tree-rats and mice possess a callosity beside the teat to which the young cling and are carried from place to place, thus escaping the peril of a permanent nest. He likewise commented on the efficacy of the marsupium of the opossums for the same purpose; but here his usual precision failed him, for in the Didelphidæ the marsupium is rudimentary. Nevertheless his contention holds good, for many neotropical opossums carry their young on their backs from place to place. It is noticeable likewise that the sloths and monkeys transport their sucklings with them, instead of building nurseries in the ant-infested trees.

Atta cephalotes, the leaf-cutting ant of Guiana, forms communities so large and numerous as to render the cultivation of certain crops almost impossible. The *Atta* parties work chiefly at night, but even by day attention is frequently drawn to them by the sight of a patch of foliage melting away before the eyes as hundreds of great red ants shear out bits of leaf and carry them away in procession down the stem. They often forage at considerable distances from the nest, and well-marked trails radiate out into the jungle from their headquarters. I have watched them ascend a forest tree sixty or seventy feet high to obtain the foliage, and then travel eighty or a hundred yards to the nest. Marching thus, each with a little green disc the size of a sixpence held aloft like a banner, an *Atta* column looks, as Belt expressed it, like a little Birnam Wood going to Dunsinane.

The cutting parties contain two forms, the large ants who do the work, and a number of small attendants who seem to take no part in labour, but who frequently enjoy a free ride on the loads of their sisters. The nest is made underground, but the surface is raised into mounds and hummocks as in a rabbit warren, and for yards round the undergrowth is stripped of its foliage. The first spadeful of soil removed in excavation brings up a horde of great soldier ants with enormous heads and great mandibles which shear right through clothing and give a painful nip. These soldiers never leave the nest and their functions are purely defensive, for the *Atta* is a peaceful agriculturist by profession. They are fungus cultivators, and each species propagates its own kind of fungus. The bits of cut leaves are carried into the nest and chewed and manipulated into a mass many inches in diameter. This heap of decomposing vegetation is infected with the mycelium of the fungus and is attended by a third caste of the ants. They clip and treat the growth in such a way that conidia never appear, and the mycelium itself develops white succulent knobs (bromatia) which are the food of the ants. The fungus can be isolated and grown apart from the ants, and in such cases it produces conidia freely. Exceptionally also a pileate sporophore like an agaric appears in deserted nests.

It is natural to ask: How do the ants obtain the fungus in the first place, since it is essential to their existence and is not known in a wild state?

The nest is founded by a single alate queen, who, after a brief nuptial flight with a short-lived mate, descends to earth, casts off her wings and makes a small chamber underground. Unlike the queen honey-bee, who is accompanied by a retinue of devoted workers, the ant begins her new life all alone. She has however brought a dowry upon which the prosperity of her future mighty brood depends. In common with other ants, she possesses a small pouch below the mouth opening which serves as a receptacle for food and other matter. When the young queen starts forth on her travels she carries in her infra-buccal pocket a pellet of the fungus hyphæ and their substratum. It is like the sacred fire which used to be handed on from one generation to the next and never allowed to die out. Arrived in her subterranean chamber, the queen casts forth her pellet, and the fungus begins to grow. She tends it solicitously, manuring it with her fæces, and it is said that she even sacrifices some of her eggs and breaks them up to fertilise her garden (51). As the eggs are laid, they are placed on the developing fungus bed, and as the larvæ hatch, they feed on the mycelium. When they grow up, they take over the garden-tending and other duties of the nest, while the queen devotes herself for the rest of her life to the production of eggs. Some idea of her enormous fecundity can be gained from the statement that a nest of the Brazilian species *Atta sexdens* may contain as many as 600,000 individuals of several different castes, workers, soldiers, fungus tenders, nurses, scavengers, etc.[1]

Besides the fungus, the queen *Atta* on her marriage flight sometimes carries other furnishing for her future home. The nests of *Atta cephalotes* are inhabited by a minute cockroach called *Attaphila*. This insect is blind, and is so modified in form and habit from association with the ants that it is inconceivable that it could find its way unaided from one nest to another. But it was observed in an excavated nest that seven out of twelve young queens ready to depart, carried *Attaphila* clinging to their bodies; and thus there is reason to believe that the queen herself transports these helpless roaches to a new home (9). Queen ants of other species sometimes carry more useful passengers. Wheeler (51) discusses the case of an Ethio-

---

[1] The huge population of some of these long-established nests is perhaps the offspring of more than one queen; for, unlike hive-bees, several queen ants may dwell amicably and reproduce in the same community.

pian species[1] in which the foundress is several thousand times as large as her workers, and so is too big to minister to her brood. Therefore when she leaves the nest, she bears, clinging to her hairy feet, a few minute workers, who, when they arrive in the new home, will attend to the eggs and larvæ and assist in the foundation of the colony.

The leaf-cutter ants are industrious and respectable horti-culturists. Communities of another kind are those of the army-ants[2]. They are predaceous hunters and robbers, who have no fixed abode but travel about the jungle, carrying their queen and their larvæ with them. As they pass, they slay and devour with a sleuth-like exactness which is all the more marvellous because they are blind. When the column comes to a crack in the ground, some individuals seize each other and swing as a living bridge across the chasm, over which the rest of the tribe pass. The breadth of the column is not very great, luckily for the mobile denizens of the forest who hasten out of the line of march; but it often takes a long time to pass a given point for the individuals composing it may be exceedingly numerous. The nest is a temporary arrangement formed in a hollow stump or similar shelter. It is one of the most remarkable nests in the animal kingdom, for it is composed of a great agglomeration of the ants themselves. Belt(10) thus describes this extraordinary nursery:

The ants were clustered together in a dense mass like a swarm of bees, hanging from the roof but reaching the ground below. Their innumerable long legs looked like brown threads binding together the mass, which must have been at least a cubic yard in bulk, and contained hundreds of thousands of individuals, although many columns were outside, some bringing in pupæ of ants, others the legs and dissected bodies of other insects. I was surprised to see in this living nest tubular passages leading down to the centre of the mass, kept open just as if it had been formed of inorganic materials.

These observations have been confirmed by others. The larvæ and pupæ and the queen are kept in the centre, and the returning hunters with the prey pass inwards through the

---

[1] *Carebara vidua.*   [2] *Eciton.*

entrances, while from the periphery fresh bands go forth to forage. In fact the ant gathering is like a river which retains its identity in continual flow. "Plus ça change, plus c'est la même chose." And, throughout, a ceaseless rain of empty insect bodies—the remains of the prey brought in—filters to the ground.

We know very little of the bionomics of these ants, or of the origin and duration of the living nest, but there are indications that it may be a temporary caravanserai, necessary to tide over the critical time when a batch of larvæ are ready to pupate. Metamorphosis in insects is accompanied by profound biological changes, and it is possible that when these are imminent, the army halts awhile to give the developing brood the opportunity to transform in peace and safety.

It is not known precisely how new bands of army-ants arise, for the queens are always wingless and have no nuptial flight. It is suggested that the helpless unwieldy creatures are sought on the ground by the alate males, and that the army then divides, one legion of workers attaching themselves to the young queen to form a new community, somewhat as a bees' nest is founded by swarming.

The army-ants are among the most successful hunters in the jungle, but their expeditions are not undertaken without risk. The bands are attended by a number of hangers-on, such as predaceous beetles, flies, and other insects, which prey on the refugees escaping from the raiders' line of march, and which, like wolves following a caravan, are not above despatching the crippled ants themselves. The white-fronted ant-thrush, according to several observers, is rarely seen except in attendance on army-ant expeditions, when it finds abundant food in the insects disturbed by the advancing raiders. The driver-ants of Africa are likewise followed by several species of birds, and certain lizards, such as the skinks, *Mabuga* sp., which prey upon the ants themselves.

Social organisation and caste differentiation are highly developed in Termitidæ. The termites are allied to the cockroaches and grasshoppers, and represent a comparatively ancient and generalised type of insect structure.

The following castes are known in termite communities:

Royalties, of both sexes and two or three forms.

Soldiers,   „   „   „

Workers,   „   „   „

Not all of these are present in the same nest, although half a dozen or even more castes have been found in a single community. The more primitive species have not developed such a high state of social organisation, and may dispense entirely with the soldier or the worker caste. The essential difference between the communities of the social Hymenoptera and the termites is the fact that in the former the soldiers and workers are females with reduced or non-functional ovaries, while in the latter they are individuals of both sexes whose development has been arrested in some way. Moreover, in the termites, the male does not perish after mating, but assists the queen to form the new nest and afterwards lives with her in the same royal chamber.

The soldiers and workers are frequently very unlike in appearance, but in some species their functions seem to overlap, and the soldiers assist in building operations. In others the soldiers have enormous mandibles for defence, and a form is known, called a "nasutus," in which the head is produced into a retort-shaped form with a gland at the apex. These nasuti, when excited, eject a sticky fluid which entangles the antennæ and legs of a stranger insect. In fact they resemble living syringes of bird-lime.

The origin of termite castes has given rise to much discussion and cannot as yet be regarded as settled. The older observers held that it was due to the kind of food given to the nymphs; but later work has tended to modify this view, for it has been shown that the sterile castes can be distinguished from the sexual forms even at hatching, and it is possible that differences may exist in the eggs. Imms (25), who discusses the matter at some length, offers an explanation based on the Mendelian Law of inheritance.

The soldiers and workers are usually sterile, though in rare cases eggs have been laid. The population is maintained by eggs laid by the queen. The termite foundress is one of the most remarkable examples of fertility known. It has been calculated that in her life of about ten years she may lay as many as 100,000,000 eggs, that is about 30,000 a day. Her abdomen becomes so distended that her vast bulk quite dwarfs the proportions of her mate, and may be 20,000 times that of her worker progeny.

The death of the king or queen from old age does not lead to the extinction of the community, for in such cases the workers are able to rear certain of the young forms to take their place. These substitutes, or "neotenic" royalties, as they are sometimes called, do not develop wings or other secondary sexual characteristics of the true royalties. They can lay and fertilise eggs, but their fecundity is relatively low, and it requires several neotenic queens to maintain the community at its full strength. Moreover the neotenes cannot give rise to the full royalties: they can only reproduce the castes similar to or inferior to their own. There is even reason to believe that in some species the winged forms no longer appear, and that the existing communities are maintained only by neotenes.

In the neotropical forests, termites compete with ants for the place of most abundant insect. Their nests are ubiquitous and of every possible size and shape. Some are diffuse and represented only by a system of passages in the soil or in rotten tree stumps; others are discrete and form globular masses, often as hard as concrete, on the branches at every level above the ground. Some forms encase an entire tree trunk and build drip-points—projections slanting downwards—to carry off the rain. But in spite of their abundance, the termites are not as conspicuous as the ants to whom they offer so many interesting analogies, because they do not work in the open, but travel by long covered galleries which they build up trees and walls and in the soil. Arboreal nests have systems of passages running down to the ground, and in some species the majority of the community at all events are subterranean and use the aerial nest only as an annexe.

Termites form the centre of an interesting bionomical complex. In the first place they are the staple food of many mammals and birds. Some burrowing snakes, such as species of *Lepto-typhlops*, live in the nests and feed on the inhabitants. Certain Meliponine and long-tongued Euglossid bees lay their eggs in the outer walls of the arboreal nests, and birds and lizards sometimes take advantage of the same situation. The nests also harbour a large number of inquiline species of other insects, an arrangement which is characteristic of many ant communities also[1]. Many of these guests are beetles, but representatives of almost every large order are known. Wheeler (51) classifies them as (*a*) predators, (*b*) parasites, (*c*) synœketes or tolerated guests, and (*d*) symphiles or true guests. The last class is of peculiar interest and many of its members have become structurally modified by the association. Certain beetle symphiles of ants are furnished with tufts of golden hair, called trichomes, which surround the orifices of special glands whose secretion is attractive to the ants. Some symphiles of termites possess analogous structures in club-shaped glandular processes called exudatoria. The hosts crowd round their guests to lick the secretions of the trichomes and exudatoria respectively, and in return tend the owners carefully, feeding them, and when necessary removing them from danger as if they were their own offspring. It has been suggested that these secretions may actually benefit the hosts—in other words that they may act as condiments for digestion, and that their producers may be regarded as living cruet-stands. Other observers have compared the host's infatuation for the body fluids of their symphiles to the drink or drug habit in the human race. Be this as it may, the relationship is almost without parallel in the animal kingdom. Wheeler (51) thus describes it in an amusing passage:

Any insect possessed of the glandular attractions I shall describe can induce the ants to adopt, feed and care for it, and thus become a member of the colony just as an attractive and apparently well-behaved foreigner can secure naturalisation and nourishment in any human community. But the procedure among ants is more striking because the foreigners are so very foreign—that is belong to such alien and heterogeneous groups. Were we to behave in an analogous manner, we should live in a truly Alice-in-Wonderland society. We

[1] About two thousand species of myrmecophiles have been described for the ants alone.

should delight in keeping porcupines, alligators and lobsters in our houses; insist on their sitting down to table with us, and feed them so solicitously with spoon victuals that our children would either perish of neglect or grow up as hopeless rhachitics.

The origin and development of social organisations like those of the ants and termites is an abstruse problem. Wheeler has brought his unrivalled knowledge to bear on the matter in some of the most delightful zoological essays ever written. He points out that the social instinct has arisen independently in no less than twenty-four groups, belonging to five different orders of insects; and in each case he traces its genesis to a lengthening of adult life to overlap the growth period of the young. The parent, when the brood remained together, thus acquired an interest in her progeny and was associated with them. The existing Embidaria perhaps represent this stage in the evolution of social life. These insects of low and probably primitive type inhabit warm countries, and some species weave communal webs on tree trunks or in hollows in the ground. I have seen a tree in Trinidad covered to a height of many feet with a mat of silk in which several generations of Embiids lived together in a complicated system of communicating tunnels. The female may live for some time after mating; and as her daughters and their progeny grow up around her, they enlarge and extend the silken tent she began. In some species the association is not merely fortuitous, but the parent is actively solicitous for her offspring. Thus Imms[24] describes an Indian Embiid which covers her brood with her body, and when removed hastens back to them.

The concentration of several generations in a restricted place of shelter created, we may suppose, difficulties of food-supply, and to meet this a division of labour arose. Some individuals (incipient worker caste) became specialised to bring food in from without, while to others was relegated the important function of reproduction in the security of the nest. Now for the first time each individual acquired a personal interest in every other: the community was no longer a crowd: it became an entity. This interest was physiological and intimate. The "feeders" ministered to the "breeders," supplying them at first no doubt with raw provisions, but later, as nutritional conditions became more complex, with a special pabulum in which the recipients absorbed the secretions of the donors' bodies with the prepared food.

Perhaps the nearest parallel that can be offered is the suckling of a young mammal by the parent; but to make it exact it must be supposed that the alimentary relationship is reciprocal, and that the mother in her turn receives from the offspring a secretion essential to her well-being. That this state of affairs exists in many insect communities admits of no doubt. Roubaud (40) and others have shown that the saliva of wasp larvæ is eagerly sought after by the workers, and the same exchange of nutriment in more marked degree is found among the different castes of termites and ants[1]. Wheeler calls this phenomenon "trophallaxis." He sees in it the unifying bond of these complex communities, and regards their relations with the symphiles as perversions of the trophallactic instinct. He even goes so far as to compare a termite colony to a vast organism whose members, though without organic continuity, are nevertheless united and mutually nourished by "a circulating medium of glandular secretions, fatty exudations, and partly and wholly digested food, just as the cells of the body of a higher animal are bound together as a syntrophic whole by means of the circulating blood."

With our present knowledge of vitamines, endocrine secretions, etc., it is permissible to suggest that the effects of these trophic inter-relations may be far-reaching. They even offer an explanation of the development, though perhaps not the determination of caste, for it is possible that though social polymorphism may be based upon germinal distinctions, its manifestation may depend upon trophic conditions[2].

However this may be, there is no doubt of the defensive value

---

[1] Wheeler remarks that it is absent in bees, but he gives a possible reason for its loss in these insects.

[2] Besides the associations mentioned here, in three out of the four families of termites the alimentary canal contains numerous symbiotic protozoa. The latter cannot exist elsewhere; and as they digest the wood which is the principal food of the insects, it has been proved that life without their unicellular partners is impossible for these wood-eating termites.

of social organisation. Ants and termites are highly successful insects from the point of view of numbers and variety of environment and mode of life. Moreover their success is no modern development. The amber beds of the Lower Oligocene show that at that distant epoch the polymorphism of ants was substantially what it is to-day, and that their structure has undergone little change. Wheeler holds that they may have arisen in the Cretaceous, or even earlier. The evolution of the ants and termites has followed two rather different paths. On the whole the ants have retained their freedom. Protected by chitinous armour, stout jaws, or poisonous secretions, most of the existing species can wander abroad on their lawful occasions; for although they possess enemies, these are not numerous enough to threaten the community[1].

With the termites it is otherwise. They have substituted artificial for natural defence, and withdrawn further and further into dug-outs and fortresses of concrete-carton. Only a few species now wander abroad in daylight. This cloistered life has inevitably led to structural degeneration in reduced eyes, thin cuticle, and stout unwieldy bodies. Just in proportion to these changes, their value as food to predators has increased, and there is no more persecuted race in the world. Birds, mammals, reptiles and amphibians tap their galleries, carnivorous beetles and bugs invade their strongholds; and their bitterest enemies of all are the ants, their rivals, who prey upon them, plunder them and occupy their cities. That they exist at all is due to their marvellous architectural industry which until now has kept pace with their persecution; but according to Wheeler there are indications that termite evolution is already past its hey-day, and that the palm of the future for social organisation and progress will fall to the more plastic and less restricted ants.

[1] See Bequaert (12).

CHAPTER IV

At the conclusion of the second chapter it was pointed out that the two principal ways in which animals escape the attacks of their enemies and hold their own in the struggle for existence are by defensive associations of individuals and species, such as have just been described, and by the development of an inconspicuous or misleading appearance which causes them to be overlooked by their foes. As neither of these forms of defence is confined to tropical countries, two chapters of an essay on rain-forest œcology may seem excessive space to devote to them; but the excuse is that in both cases, and especially the latter, as the favourable conditions of the tropical jungle support an exceptionally large and varied population, so the struggle for life between the eaters and the eaten is exceptionally severe there and has given rise to an infinite diversity of devices for concealment. Sometimes these are for offence, as when the predatory mantis, shaped and tinted like a flower, lurks among flowers for the insects which come to sip nectar. Sometimes they are for defence, as when the soft-bodied caterpillar mimics the green-veined leaf on which it grazes and so evades the keen eyes of hungry birds. Both these kinds of deceptive appearance are instances of what is usually called protective coloration, and the principle of this is held by most naturalists, although opinions differ as to how the various devices have arisen and how far many of them really serve their purpose. Hence to avoid repetition it seems best here merely to give a brief outline of the subject and then approach it from the angle of personal observation; for experience has convinced me that explanations of colour pattern based on the examination of museum specimens may be misleading unless checked by study in the field.

On his first visit to tropical collecting grounds, the naturalist passes through four stages of reaction towards protective coloration. For the first few days he is a humiliated dupe. Wasps of standard pattern turn into harmless moths in his net; bird droppings come to life and scuttle off as spiders; and green leaves take wings and skip away as grasshoppers. The next stage is one of suspicion. He stalks bits of lichen under the delusion that they are cicadas and weevils; he captures a spider with precaution, only to find that the nimble Arachnid is nothing but a sedentary caterpillar; and he constantly sweeps up common ants and bugs in the search for yet more wonderful impersonations. The third stage is reassuring. Our naturalist believes that the scales have fallen from his eyes; his collection grows, and he wonders how he was ever deceived by such transparent devices. But sooner or later he enters on the fourth period when transparent devices no longer deceive him, but, he asks himself uneasily, how he is to determine whether a device is transparent. In the Guiana forest there is a whole group of large parasitic Hymenoptera with banded semi-hyaline wings and orange-red bodies. In his noviciate, our collector nets what is apparently one of these, and finds that it is a predatory Reduviid bug. Delighted with this example of mimicry, he takes every specimen he sees, and catches nothing but Hymenoptera. By and by he learns that there are certain characteristics in the carriage of the wings which distinguish the mimic from the model, and in this belief he passes creditably through the third stage. Then one day he sees an Ichneumonid, which, he considers, a less experienced collector might have mistaken for a bug. He captures it and behold—it *is* a bug—not a carnivorous Reduviid, but a vegetarian Capsid. Thus are simple, double, and triple bluffs "put over" the entomologist. It is a question how far they are also put over the predaceous enemies of insects, and which of the four educational phases of the human collector is that of a hungry bird. In an inquiry of this kind, as it is possible to apply only the human criterion it is usually assumed that the lower animals see as we do. For the invertebrate eye at least this is a large assumption to make, though probably the vision of a bird is as acute as that of man, if not more so.

Professor Poulton (34) has devised a classification to cover all kinds of animal coloration, which, though convenient, is too elaborate for our present purpose. It includes however four classes of coloration, which, as they are particularly related to defence against attack, may be quoted here.

I. Resemblance for defence to some part of the environment (usually inanimate) or Procryptic Coloration.

II. Conspicuous colours which give warning of unpalatability or Aposematic Coloration.

III. The possession in common of warning colours and appearance by two or more unpalatable species or Synaposematic Coloration, sometimes called Müllerian Mimicry.

IV. The possession of false warning colours by a defenceless species or Pseudaposematic Coloration, sometimes called Batesian Mimicry.

The first (procryptic) class is perhaps the least controvertible. The fact that the undersides of the wings of certain butterflies resemble dead leaves, and that some caterpillars look like twigs, is a matter of common experience for the entomologist. The wonder is that the same natural groups should achieve obliteration in so many different ways. In a few yards of South American jungle the most diverse kinds of grasshoppers blend harmoniously with various surroundings. Dark green Tryxalidæ squat upon the upper surface of aspidistra leaves, the lines of their extended bodies and antennæ following the ribbed venation; and on an adjoining tree, little crickets, brown and wrinkled, pass for nodules of bark. A long-horned grasshopper presses flat against an upright stem, and his body bears a pectinate fringe which clasps his perch and obscures his outline. His colour alone seems to have no protective purpose, for it is sand-grey and stands out with some distinctness against the dark background. A second glance makes its meaning clear. The grasshopper simulates, not part of the twig to which he clings, but the empty cocoon of a moth or the dry egg-case of a mantis.

The heteropterous bugs offer excellent examples of diversity of form to attain the same end. A gregarious Pentatomid[1] clusters on the stems of lianas in deep shade, and passes as an irregularity of the bark. Another bark-living form, the Aradid *Dysodius lunatus*, is greatly flattened dorso-ventrally and the lateral margins of the body are deeply serrate. These two examples of procryptic coloration are of interest because both

[1] *Cytocoris gibbus.*

belong to families whose members when handled emit a nauseous fluid with a strong repellent odour. The difference in form between the globose Pentatomid and the flattened Reduviid has a bionomical purpose. The former is a sedentary vegetarian and can afford to imitate a nodule of bark. The latter is a mobile hunter of other insects and if rotund would attract attention at once. Instead, it is flattened like a flake and glides unnoticed over the bole of the tree. The bugs of the subfamily Emesinæ are likewise procryptic; but in contrast to *Dysodius*, whose haunts they share, they achieve their end by excessive elongation of the body and limbs which are of threadlike tenuity.

As a rule phytophagous insects, such as the plant-sucking bugs, are sedentary; and their colour pattern, when protective, simulates vegetable structures like bark, leaves, or seeds, while resemblance to other animals is less frequent. With some exceptions, the converse holds good for the predatory Reduviidæ, since the disguise of an inanimate object would soon be betrayed by the restless movements of the hunter.

While many instances of procryptic coloration admit of ready and probably just interpretation, others are doubtful. It is not enough that the resemblance in appearance and behaviour impresses the human observer. There should be evidence that the enemies are deceived, and that the subject itself habitually occupies an environment to suit its disguise. I came across an instance of this in Guiana. The purple and yellow chrysalis of a common butterfly[1] bore a superficial resemblance in form and colour to the tubular flowers of a catchfly vine which was abundant in the forest. I made particular search to ascertain whether, and if so how frequently, the pupa was attached to the plant itself; but although both occurred together in the same small area, I was not able to convince myself that the resemblance was more than fortuitous. The naturalist in the tropics must be ever alive to the danger of over-riding a fascinating theory.

A bug[2] from the same place offered a more satisfactory conclusion. This insect was green and elongated, variegated with

---

[1] *Papilio æneas.*   [2] *Leptocorissa tipuloides.*

purple and red. It fed, literally in hundreds, on the panicles of a sedge-like grass which grew in patches in dry clearings, and was never taken in any other situation. Its likeness to the long green and purple glumes was so pronounced, and its environmental range in this area so restricted, that probably it may be safely cited as an example of protective resemblance.

Several insects in different parts of the world disguise themselves as bird droppings. These cases are of interest since the resemblance is sometimes for offensive and sometimes for defensive purposes, and may be effective against one enemy and not against another. Thus birds and lizards are probably deceived; but many insects, such as butterflies, flies, ants and bugs, seek out excrement for the sake of the moisture[1]. Hence it is justifiable to suppose that excrement mimics are in danger from birds but not from sucking insects.

Several caterpillars resemble bird droppings. I remember collecting half a dozen larvæ of a large swallowtail butterfly on a vine in British Guiana and bringing them into the house for examination. A few minutes later, I saw what I supposed to be one of these caterpillars accidently dropped, lying on the doorstep, and picked it up. However it turned out that I was quite deceived, and that the object was really excrement, dropped by a captive trumpeter tethered near the house. A small moth[2] taken near the same place also bore a striking resemblance to a bird dropping (Pl. IV, fig. 3).

Forbes (21) describes an East Indian spider which evidently belongs to a class called by Poulton "pseudepisematic"—that is, the colour pattern of a predatory species, designed to lure prey within reach. This spider is black and white, and as it lies on its back on a leaf, it resembles a lump of excrement. To complete the likeness, a thick web is woven below it in such a way as to simulate the liquid draining over the leaf. This alluring device is sometimes successful, for Forbes saw the spider catch a butterfly which alighted to sip the supposed fluid.

[1] Some neotropical Coreid bugs are losing their original sap-sucking habits and live largely on nectar, honey-dew and the juices exuding from excrement and decaying matter. From sucking the dead, it is but a step to the living—a step which the family Reduviidæ has already taken.

[2] *Stenoma* sp.

Warning coloration is more difficult to determine than cryptic coloration, because while observation decides that such and such an animal resembles a broken twig or a dead leaf, inference and experiment are required to judge whether a particular type of colour pattern warns off enemies. There is a good deal of evidence that brightly coloured insects are as a rule rejected by birds, monkeys, lizards, etc., because these enemies have learned individually that a gaudy colour pattern is often associated with stings, poisonous hairs, an offensive smell or an acrid taste. The experiments of Finn and Swynnerton on captive vertebrates bear out this view. It should however be pointed out that much of the success of colour advertisement depends on the appetite of the enemy. As it has been put: A traveller finds a packet of sandwiches left on the seat of a third-class railway carriage. If he is starving he will probably devour them at once. If not, he may be fastidious and prefer to wait arrival at a station buffet instead of satisfying his hunger with provisions from an unknown source. Moreover, an animal may have more than one kind of enemy, and a defence efficacious against one may not hold good for another.

Bright colours alone do not necessarily indicate unpalatableness. For example, certain mantids and parrots, which are highly conspicuous in a glass case, may be quite obliterative against their natural background of sunlit foliage. The best criterion for a suspected case is not only whether the subject appears conspicuous in its habitual surroundings, but whether it is common and sits about in plain view. This is the case with many of the bugs, such as the Fulgoridæ, which secrete masses of flocculent white wax, and the Heteroptera, which are known to possess stink-glands.

Procryptic and warning coloration are sometimes combined in one individual. The caterpillar of a certain South American Sphingid moth is pale brown, and in its attitude of repose resembles a dead rolled leaf both in form and colour. But when it is irritated or alarmed, the larva suddenly raises the anterior third of the body and displays on the ventral surface two pairs of conspicuous yellow eyespots with black centres which were previously concealed from sight in the inter-segmental folds.

DESCRIPTION OF PLATE III.

Protective coloration in some neotropical bugs. × $\frac{7}{8}$.

1, Müllerian association between: *a*, *Scaphura nigra*; *b*, *Spiniger spinidorsis*; *c*, *Pepsis chrysobapta* (model). 2, Müllerian association between: *a*, *Notocyrtus gibbus*; *b*, a Meliponine bee (model); *c*, *Teletusa* sp. 3, *Spiniger* nov.sp. 4, *Umbonia spinosa*. 5, dorsal and lateral aspects of *Heteronotus armatus*. 6, *Cyphonia clavata*. 7, *Membracis c-album*. 8, aposematic coloration in Homoptera: *a*, *Darnis partita*; *b*, *Lystra lanata*; *c*, *Tomaspis ruber* (Cercopidæ). 9, Müllerian association, with white antennæ: *a*, the Ichneumonid model; *b*, *Hyalymenus* sp.; *c*, *Holymenia histrio*; *d*, *Calobata* sp.; *e*, a yellow and black roach. 10, Müllerian association between: *a* and *d*, Ichneumonid models; *b* and *c*, *Xystonyttus nigriceps* and *X. nugax* (Reduviidæ); *e*, *Monalonion* sp. (Capsidæ). 11, procryptic Membracidæ: *a*, *Lycoderes hippocampus*; *b*, *Amastris vismiæ*; *c*, ♀, and *d*, ♂ of *Stegaspis galeata*. 12, nymph of *Hyalymenus pulcher*. 13, Müllerian association between: *a*, Pompilid wasp (model); *b*, *Pompiliopsis tarsalis*; *c*, *Spiniger nigripennis*.

PLATE III

PROTECTIVE COLORATION IN SOME NEOTROPICAL BUGS

The transformation is quite startling enough to arrest the hand of a human collector (Pl. IV, figs. 2, 2 A).

Poulton (37) has made the interesting suggestion that in countries with a well-marked wet and dry season, insects with aposematic colouring are relatively less abundant in the dry season. It is suggested that this is because the stress of existence is greater in the dry period. All life is then less abundant, and as fewer palatable forms are available, predatory enemies are obliged to satisfy their hunger on food which in times of plenty they pass by with distaste. Hence in the dry season, aposematic coloration is dangerous rather than advantageous to its possessors, and a proportionately large number of the species that remain active are procryptically coloured. It would be of interest to test this theory in the rain-forest, but unfortunately necessary data are lacking. The climate of the Guiana forest differs from that of a region where rain and comparative drought alternate, in that it is always humid, and the wet season of May and June means less the coming of rain to a dry country than the withdrawal of sunshine from one that is already saturated. The time of renewal of foliage and fresh activity is in the latter part of July and early August, when the bright intervals are prolonged. As Bates remarked of the campos near Santarem: "The heaviest rains fall in April, May and June; they come in a succession of showers with sunny gleamy weather in the intervals. June and July are the months when the leafy luxuriance of the campos and the activity of life are at their height." It is true that Howes (23) found that the incidence of nesting of the fossorial Hymenoptera in Guiana rose steadily from February till June in the height of the rains; but what is true of one group may not hold good for all, and it is to be hoped that the considerable amount of data obtained at the tropical laboratory of the New York Zoological Park at Kartabo will ultimately throw some light on the matter. Meanwhile, with a view to testing the theory of season and aposematic coloration, I prepared curves of my own captures of three species of solitary conspicuous bugs at Kartabo between June and September 1922. *Lystra lanata* is a black and white and scarlet Fulgorid, which secretes a mass of white flocculent wax at the hinder end of the body. The curve of this species is not altogether satisfactory as I did not collect strictly every individual seen in August and September; but it is included here because the large increase in August is confirmed by a note in my journal at the beginning of the month (Pl. III, fig. 8 B).

Of the other two species, every example seen was taken and dated. *Brachystethus cribrum* (Pentatomidæ) and *Nematopus indus* (Coreidæ) are both large conspicuous black and red bugs with a powerful odour. They are found on open foliage and

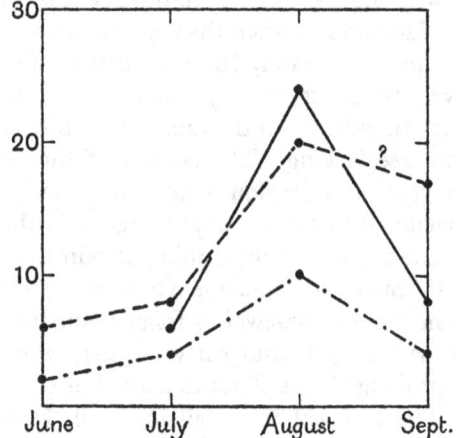

Fig. 5. Curves showing the occurrence of three conspicuous bugs at the close of the rainy season.

——————— *Nematopus indus*
—·—·—·— *Brachystethus cribrum*
— — — — — *Lystra lanata*

readily take wing. The two curves coincide in a remarkable manner. Although the decline in September is as yet unexplained, there is no doubt that these gaudy species appeared in the greatest abundance after the rains, that is to say at a period of renewed activity and increased population (Fig. 5).

Pseudaposematic coloration was first described by Bates[3] in a classic paper on the butterflies of the Amazons valley. He pointed out that the colour pattern of certain species departed from that typical of their family or group and resembled that of some gaudy (aposematic) forms to which they were only distantly related. He suggested that the former were unprotected and liable to attack, and that it was to their advantage to resemble distasteful forms, especially those more numerous in the region than themselves, since predatory enemies recognising the colour pattern would mistake them for their models and leave them alone. Thus some of the Pieridæ, which according to Bates are eaten by birds, have lost the colours and patterns typical of their family, and assumed those of certain Heliconiidæ which are known to be distasteful. This is the so-called "Batesian mimicry," for which Wallace suggested the following criteria:

(a) That the imitator species occurs in the same area and environment as the imitated.

(b) That the imitators are always the more defenceless.

(c) That the imitators are always less numerous in individuals.

(d) That the imitators differ from the bulk of their allies.

(e) That the imitation is external and visible only, and not due to affinity.

But Bates himself recognised that there were facts which could not be squared with his theory, for in some cases not only was the mimic actually more abundant than the model, but the models themselves imitated one another. In 1879, Fritz Müller[30] offered an explanation. He suggested that, given two unpalatable species, a common colour pattern would be to the advantage of both, inasmuch as it would halve the loss to both forms during the education of their enemies. Müller held, and subsequent experiment has confirmed his view, that birds have no innate knowledge of what is good or bad to eat, but each individual must gain its own experience. If the two aposematic species of different types inhabit the same area, each must sacrifice a certain number of its members, say six per cent., to the education of vertebrates. But if both forms possess the same kind of warning coloration, the birds need learn to recognise one set of signals only, instead of two, and the loss to each insect species, during the period of trial and error, will be halved. This is the theory of the so-called

"Müllerian mimicry," called by Poulton "synaposematic coloration," and there is increasing evidence that many, if indeed not all, of the instances hitherto attributed to Batesian mimicry really belong here.

It is clear, as Poulton has pointed out, that Müllerian mimicry is not mimicry in the strict sense of the word, but is really the possession of certain warning colours in common; and Dixey (20) has even produced evidence to show that in some cases the resemblance is reciprocal, each of the protected species tending to approach the type of the other. Some of the Müllerian mimicry associations include species from the most diverse groups, some more and some less effectively disguised. However, from the point of view of protection, the larger the association the wider the advertisement, and the better for the participant forms.

In the present state of our knowledge it is often difficult to determine whether we are dealing with Batesian or Müllerian cases. For instance, a certain neotropical moth[1] resembles a Lampyrid beetle, one of a group which is supposed to be distasteful to vertebrates (10). The pale narrow wings are folded back in a way reminiscent of the hard curved elytra of Coleoptera, the colour and attitude when at rest are beetle-like, and even the edge of the shield-like prothorax is suggested by markings at the base of the wings. Here the case turns on whether or not the moth itself is nauseous to its enemies. If the former it would be a Müllerian, if the latter a Batesian mimic (Pl. IV, fig. 5).

Among the Müllerian associations there are certain outstanding models which set the standard for other and probably less well-protected species. The Hymenoptera are a case in point, and within the order it is possible to distinguish several secondary groups. The five following are notable in Guiana, and no doubt further search in the same region would multiply them.

The wasp models, varied in form and specifically numerous, agree in three particulars—black, yellow or reddish colouring, active restless movements, and noisy buzzing flight. All are

[1] *Automolis diluta.*

protected by stings and are probably the least persecuted insects in the tropics. The great hunting wasps of the family Pompilidæ especially are conspicuous metallic-coloured insects with tawny or black wings and antennæ. These Hymenoptera are mimicked by various predatory bugs of the genus *Spiniger*. The resemblance is astonishing, and extends not only to the colour but to the form of the mimic, for the bug's abdomen is pointed at the apex and constricted to resemble the wasp's "waist." In *Spiniger spinidorsis* the wings are orange and semi-hyaline, and the first joints of the antennæ are orange in imitation of the shorter yellow antennæ of the wasp, while the distal third is black, filamentous and almost invisible. Another bug, *S. nignipennis*, is black with orange antennæ and mimics another common black Pompilid. Yet a third member of this genus is orange-yellow and passes for one or other of the yellow hunting wasps of the region. In my experience, the behaviour of the bugs does not altogether bear out the imitation, for they are slower in movement than wasps, and the wings are folded flat on the back instead of being carried upright. My instinctive thought when I netted *S. spinidorsis* for the first time was: "A wasp, but there is something the matter with it." The orange *Spiniger* was specially interesting because it frequented open sandy places which were just the spots visited by its models. It should be remarked that these Reduviids are so much less common than their models as to appear at first sight like an instance of Batesian mimicry, but the association is almost certainly Müllerian for the bugs are already highly protected by stink-glands. Some grasshoppers mimic these forest wasps—sheep in wolves' clothing. The remarkable case of *Scaphura nigra* has been described by Bates (4) and Poulton (38). The grass-hopper is black with semi-opaque orange wings, and the antennæ are thickened and yellow at the base while the distal portion is black and filamentous, so that when they are laid back in the resting position only the yellow portion is visible. In spite of the fact that the carriage of the wings is different in model and mimic, the appearance of the grasshopper is truly formidable. Certain other black Pompilidæ are mimicked by Syntomid moths. The wings are semi-hyaline and the abdomen is petiolate. I caught one of these moths[1] in the Botanical Gardens at Georgetown in 1922; and the resemblance extended even to the behaviour, for the moth was fluttering over some open grass after the manner of the spider-hunting wasps of the same place (Pl. III, figs. 1, 3 and 13).

The little social bees of the family Meliponinæ are the focus of another mimicry association. These insects are stingless and at first I was at a loss to account for the advantage to their mimics. However, Wheeler(48) has recorded that a Central American species can emit a fluid caustic enough to burn the human epidermis, and it is quite possible that other forms also possess this power. Two remarkable mimics of these bees are found in the same forest clearings of Guiana, and both are bugs (Rhynchota) belonging to different divisions of that great order (Pl. III, fig. 2). One is a Jassid[2] which feeds on cassia flowers. It is about the size of a bee which is common in the locality; the wings are transparent; the body is dark brown and hairy; and the abdomen, which terminates in a pointed ovipositor, is inflated and marked with orange. The other mimics are Reduviidæ of the genus *Notocyrtus*. The head is small and inconspicuous in life, while the thorax is much inflated and divided by lateral grooves into two parts which give a fair resemblance to the thorax and square head of the bee. Moreover, the hind tibiæ are flattened and expanded into the semblance of the "pollen baskets" of the bees. The bugs live on open foliage and may be found sipping from extra-floral nectaries. The Meliponids visit the same places for the sake of nectar or wax and resin, and it is not uncommon to see model and mimic side by side. The movements of the bug are quite bee-like, and I have been deceived more than once by its appearance. Notocyrtids fall into two groups. One is dark coloured, the other is yellow and black. In the district round Kartabo where I observed these forms, the dominant Meliponinæ are black and the yellow species are scarce, but both kinds of Reduviid mimics are found. However, in Müllerian mimicry, the relative abundance of models and mimics is immaterial, for when an advertisement of unpalatability is once patented, all its wearers

[1] *Pompilopsis tarsalis.*   [2] *Teletusa* sp.

participate in the benefit. In Central America the yellow Meliponinæ are mimicked by certain Longicorn beetles, whose elytra are rudimentary and display the hyaline underwings and the hind legs bear an arrangement of hairs which simulates the "pollen baskets" (13) (Pl. IV, fig. 6).

The sawflies and parasitic Ichneumonoidea (Hymenoptera) are both considered to be distasteful groups. Certain Syntomid moths mimic sawflies and also other families of Hymenoptera. A few likewise take Diptera for models[1], and here the mottled wings of a biting fly are reproduced, not by transparency, as in the hymenopterous mimics, but by the arrangement and shading of the scales themselves.

The most notable Ichneumonid mimics are certain Reduviid bugs which reproduce admirably the banded wings and slender orange bodies of their models. Similar mimics have arisen independently among the vegetarian Capsidæ also[2]. From the mimicry standpoint the Heteroptera have been less studied than other groups, but they are second to none in interest. Müllerian mimics are most numerous among the Reduviidæ, while the plant-feeding families more often exhibit simple warning colours, a point difficult to explain since both possess defensive stink-glands. If the Reduviidæ chose innocuous models, it might be suggested that the resemblance facilitated their approach to their victims, but on the contrary they imitate highly aggressive insects. It might benefit a skunk to appear like a sheep, but why should he adopt the disguise of a tiger? (Pl. III, fig. 10.)

Not only is Müllerian mimicry to the advantage of the participating species, but one association may benefit by its resemblance to another, particularly when the type models are akin. Thus the Ichneumonid-Reduviid association just described is allied to another which has been built up round other Ichneumonids and includes a heterogeneous collection of insects. The models are large Ichneumonids with conspicuous white-tipped antennæ which vibrate rapidly as the insects run over the foliage in search of prey. Their mimics include two Coreid bugs, a cockroach and a fly, all of which have white-tipped

---

[1] E.g. *Eucereon metoidesis*.   [2] E.g. *Monalonion*.

antennæ and haunt the same places as the Hymenoptera. The bugs, which are active nectar-feeding and saprophytic forms, run with quick nervous movements of the wings and antennæ. The cockroach, though coloured black and yellow, is a comparatively poor mimic, and his disguise is soon penetrated. The fly[1] is a remarkable case. The minute antennæ are replaced in the mimicry make-up by the anterior feet which are white, and in the resting position gently wave to and fro. This association, as compared with the Ichneumonid-Reduviid group with which it really blends in some particulars, is dynamic rather than static, depending on mimicry of behaviour rather than of form. Similarity of structure is secondary, or the fly and the cockroach could scarcely participate. Success is due to the reproduction of an outstanding characteristic of the models—the vibrating white antennæ (Pl. III, fig. 9).

Ants on the whole are a distasteful group, though from the researches of Bequaert(12) it appears that their enemies are more numerous than is generally supposed. In different parts of the world they are preyed upon by various spiders, beetles, and Hymenoptera. Numerous lizards are known to devour them, and they have given their name to a whole family of neotropical birds (Formicariidæ). Amphibians are also among their enemies. Bequaert examined 194 toads belonging to five species from the Congo region, and found that ants of seventy-two different forms constituted fifty per cent. or more of the stomach contents. It is notable that these ants were almost all terrestrial forms—added proof of the advantages of arboreal life in the jungle. Several neotropical batrachians, such as the giant toad, *Bufo marinus*, are also known to eat ants. The great ant-eater is really a termite feeder, and so probably is the tamandua, but the little *Cyclopes* is said to feed exclusively on tree ants(15).

Nevertheless the ant type is the focus of several mimetic associations, and spiders, grasshoppers, bugs, beetles and caterpillars, have been recorded as mimics. The nymph of a certain neotropical bug[2] bears a remarkable likeness to some large rufous species of ants which roam over the foliage in

---

[1] *Calobata* sp.   [2] *Hyalymenus pulcher.*

the forest, visiting colonies of Homoptera for the sake of their honey-dew. The petiolated abdomen, the proportion and carriage of the legs, the spines on the thorax characteristic of many ants, and even a red patch in the anal region, are faithfully reproduced. The adult is non-mimetic, and at the last ecdysis it is remarkable to see the enlarged hind legs, strong horny wings and typical bug's head and thorax, emerge from the ant-like mould (Pl. III, fig. 12).

Some spiders are mimetic of insects or inanimate objects, but others provide models for less protected forms. The same end is attained by different means by two remarkable caterpillars from Demerara. The first is the larva of a Notodontid moth, which in repose resembles a dry twig. But when alarmed, the extremities of the body are raised, the head and thoracic legs are thrown back, and between the latter two pairs of turgid purple processes (osmeteria) are thrust out. The effect is as if a brightly coloured spider of unpalatable, not to say formidable, appearance leaped suddenly upon the twig (Pl. IV, figs. 1, 1 A).

The second case is that of the caterpillar of a Limacodid moth. Here the resemblance is permanent, and the larva feeds in full view on the upper surfaces of leaves. The body is thickly hairy and provided with several pairs of lateral processes, which are arranged like the legs and jaws of the spider. The body markings are suggestive of two pairs of dorsal eyes; but the resemblance is general rather than specific, for at ecdysis the colour, though not the pattern, changes from brown and buff to grey and red (Pl. IV, fig. 4).

Mimicry of vertebrates by invertebrates is uncommon, but some cases are known. The larva of a certain South American butterfly is a case in point[1]. The caterpillar, which lives on foliage, is leaf-green with faint veined markings. On either side of the thorax is a raised spot with little opalescent facets set in a darker field. The head, which is sharply constricted from the body, is provided with two horn-like processes. The general appearance is procryptic, but when startled the caterpillar

---

[1] *Prepona* sp. (Nymphalinæ). A fuller description of this and the two preceding species appeared in the *Transactions of the Entomological Society of London*, 1926.

retracts the head and inflates the thorax. The spots then stand out from the surface like staring eyeballs, and, probably by refraction, seem to gleam brightly. The transformation is most impressive, and the effect when the larva is half concealed in foliage is that of the head of a snake or lizard with open mouth and shining eyes. Possibly this is an example of Batesian mimicry (Pl. IV, fig. 7).

The Membracid bugs provide an instructive study in protective coloration, for within this natural group every kind of procryptic, aposematic and mimetic form and pattern is found. The body and limbs of the insects play little part in the disguise, which is determined by the bizarre development of the prothorax which extends over the body like a hood. Systematists divide the Membracidæ into five sub-families, and it appears that the same protective devices have arisen independently in several cases. The majority of these are procryptic, and the insects resemble in form and colour the buds, thorns and leaf-stipules of the plants on which they feed. Green and brown Membracids almost invariably feed on green and brown stems respectively, never vice versa. The resemblances are often very subtle. Thus in one solitary form[1] the pronotum is black and produced into a long horn, while below it the wings are hyaline and allow the green abdomen to be seen by transparency. The insect feeds in the axils of leaves and is the perfect likeness of a withered and partly skeletonised stipule (Pl. III, fig. 11 A).

Warning coloration is uncommon in Membracidæ, but some species are black or brown with yellow or white markings, and it is significant that, according to my observation, these forms feed in conspicuous places, for instance, *Darnis partita* (Pl. III, fig. 8 A).

The Membracidæ attain their most wonderful development as mimics of Hymenoptera. Two sub-families have produced ant-like forms. *Cyphonia clavata* (Smiliinæ) feeds on leaves in open places and is of a wary and active disposition. The body and limbs, pale and inconspicuous, are completely covered by the black and hairy pronotum, which terminates behind in two spiked balls, each resembling the abdomen of an ant. In life the

[1] *Lycoderes hippocampus.*

general aspect of this Membracid is very formicarian, and it is readily confounded with the ants that run over the foliage around it (28). Seen in the cabinet, however, its mimicry is less exact than that of *Heteronotus trinodosus* (Darninæ), whose pronotum resembles a single ant with petiolate abdomen and thoracic spines. Nevertheless *Cyphonia* is so deceptive with its incomplete double representation that I have been tempted to wonder whether it is not in some ways the more specialised of the two, simulating not a single ant but a *tableau vivant* of daily life—two ants, side by side, feeding on a bit of decaying matter or a drop of honey-dew, with the pale body and wings of the Membracid to represent the food mass (Pl. III, fig. 6).

In other *Heteronotus*, such as *H.armatus*, the prothoracic mask is black and yellow, and shaped like the stalked abdomen of a wasp. These Membracids are lively and active, and frequent sunny open places. They take wing readily and fly with a noisy buzzing sound. The "thorax" and "abdomen" of the mimic wasp are sharply spined, and it is possible that the bug, like its model, is avoided by enemies (Pl. III, fig. 5).

The members of the genus *Lophyraspis* are ant mimics by behaviour only, for there is nothing ant-like in their appearance except size, colour, and the long curved hind legs. These little Membracids live in colonies on jungle plants, and as they feed they move their legs up and down in rhythmical unison. The casual passer-by mistakes them for a cluster of ants with vibrating limbs, and instinctively avoids them. As a matter of fact, *Lophyraspis*, like so many Homoptera, is regularly visited by ants for the sake of its excreted honey-dew, and when disturbed the ants swarm on to the intruder in defence of their protégés.

The value of ant protection to Homoptera is considerable, and, to realise it, it is necessary to see the abundance of ants in the tropics and the avidity with which they seek out moisture of all kinds. A decomposing caterpillar or a bird dropping brings them flocking round, but a favourite food is the excrement of Coccids, Membracids and other sedentary sucking insects. They attend such colonies constantly, and caress them into yielding up their sweets. In fact, from this point of view, the jungle may be regarded as a region with a rich but inaccessible water supply, and a huge thirsty population who attend to millions of tiny force pumps. The saccharine sap is the water, and the pumps are hosts of leaf-hoppers, frog-hoppers, Jassids and Coccids, which continually draw it up into their bodies and strew it over the leaves like manna for the benefit of ants, bees, wasps, butterflies, bugs, etc. The trophic relations between bugs and ants have perhaps a further bionomical significance. While collecting in Guiana, I commented in my journal on the comparative scarcity of the minute parasitic Hymenoptera which are so characteristic of an English woodside in summer. Moreover it was remarked that while caterpillars and grasshoppers were frequently parasitised, the incidence of parasitism among Membracidæ and Jassidæ did not seem high. It is possible that by constantly handling and caressing their sap-sucking flocks, the ants indirectly protect them from the oviposition of parasites.

The value of ants to the Homoptera is brought out by a colour classification of forty species of Membracidæ collected in one small area of forest and clearing in British Guiana. The collection was divided into two groups according to whether or not the species in question was regularly attended by ants. Each group was then further separated into procryptic, mimetic, and "neutral" forms, the last including species which, by uniform green or brown colour or small size, were inconspicuous without definite imitation of environmental objects. The results are given in the table below:

	Attended by ants	Not attended by ants
Procryptic	16	4
Mimetic	3 (behaviour)	4 (structure)
Neutral	11	0

The majority of the ant-attended forms are gregarious, while the rest are solitary. The first class contains no structural mimics—the three species included belong to the genus *Lophyraspis* and are mimetic only by behaviour. The proportionately high number of "neutral" forms in the first, and of structural mimics in the second class suggests that the attendance of ants is a defence and compensates for the lack of special protective devices, while the unattended forms must rely to a greater extent on mimetic resemblance. Two yellow and black (aposematic) species of *Darnis* are not included here, but it is significant that neither is attended by ants.

In the interpretation of colour problems it is most important that deductions made in the museum should be checked by field observations. In 1903 Poulton (36) contributed an interesting introductory essay to Buckton's "Monograph of the Membracidæ," in which he sought to assign protective meaning to various forms. In 1922 I collected some of these insects in Guiana, and it was interesting to compare the field notes with the museum inferences. Poulton justly remarks of the curious little Membracids of the genus *Bolbonota* that they "closely resemble seeds, also small lumps of earth. They would be well concealed on rough bark." The deduction that these insects resemble bark is quite correct, but the disguise is supported in an unexpected manner. Far from resting on tree trunks, they feed in conspicuous positions on aspidistras and other broad-leaved plants, and against the green background they are visible many feet away. But they simulate bark nevertheless, for the whole jungle floor is strewn with debris from the tree-tops; and bits of twigs and lichen, withered leaves and bud scales, lie so thickly on the foliage at ground level that the Membracids, though visible, pass unrecognised. Their breeding habits are peculiar. The nest is a crescentic structure made of a white waxy substance in which the eggs are embedded, and several are built in a cluster against a twig. The female sits in a cavity on the upper surface of the nest, which is several times larger than her body; and the whole colony is easily mistaken for one of the crops of small white fungi sprinkled with bits of bark which are common objects among the undergrowth.

Poulton cited a remarkable Membracid described by Sclater (35) which resembled a leaf-cutter ant bearing aloft its piece of cut leaf, and sought to extend this interpretation to the genus *Membracis* with its semi-circular hood. It is true that the ground colour in these forms is black, but to overcome this difficulty it was pointed out that leaf-cutter ants have been known to carry off bits of butterfly wings and other dark-coloured objects. From observations in Guiana, I would suggest rather that the coloration is a warning. Mottram (29) claims to have shown experimentally that black and white are the most conspicuous colours in nature, and that they are most effective when super-imposed in circular shape. The arrangement in *M. c-album* is, with modifications, typical of the genus and approaches this pattern (Pl. III, fig. 7). Moreover the colonies of this species are conspicuous objects. The nymphs are large, hirsute, and covered with a white mealy coat interspersed with black spots; and it has occurred to me that the white spotted hood of the adult *Membracis* is well adapted to blend with the mass of its conspicuous and unpalatable young. It is, in fact, a kind of group mimicry.

But there are other forms which neither field nor museum observations can interpret, in which the pronotal hood is con-torted into the most grotesque angles, bulbs, and spines to no apparent purpose. Faced with these, the protagonist of pro-tective-resemblance-evolved-by-natural-selection can only urge our ignorance of bionomics and the possibility that these forms harmonise with some part of the environment hitherto un-recognised. For the theory of natural selection offers by far the most reasonable explanation of the evolution of protective resemblance. In fact first-rate authorities have cited these re-semblances as the best evidence of the operation of natural selection. Nevertheless there are three difficulties in the way of accepting this view.

First, the lack of evidence for sufficiently strict selection by predatory enemies.

Secondly, the difficulty in accounting for the initial stages in the evolution.

Thirdly, the difficulty in accounting for the final perfecting of the protective device.

The lack of evidence of the selection of insect prey by verte-brates has been admitted by the most enthusiastic exponents of the theory; but, thanks to the labours of Poulton and others, a considerable number of observations have been recorded which show for example that birds attack butterflies more frequently than used to be supposed, and that they discriminate between the different colour patterns. An important obstacle to the theory will be removed when this evidence is amplified and extended to other groups. Thus the Membracidæ are a family with highly developed protective devices, and yet there is little or no evidence to show that they are often attacked by birds. Data from the tropics are lacking. Widermuth (52), in the United States, reports that of thirty-one birds, representing eight different species, ten had adult *Stictocephala festina* in their crops; but Funkhouser (22), from observations extending over a long period, came to the conclusion that in the Cayuga Lake region "birds are of little importance as Membracid enemies." A few species have been found in the stomachs of toads, but the latter are often nocturnal, and protective coloration is of little avail in the dark. Spiders seem to be their principal enemies, but whether the arachnid vision is comparable to that of vertebrates is at least arguable. In Guiana, I found that of three species captured by spiders, two were procryptically coloured.

It is easy to misinterpret defences against insect predators. Belt relates that he saw a green leaf-like locust stand untouched in a stream of army-ants (10), and he remarks that "the other senses, which in the *Ecitons* appear to be more acute than that of sight, must have been completely deceived." It is quite possible that the ants recognised the nature of the locust, but rejected it as food. Thus Swynnerton (45) found that the driver-ants of Africa do not slay indiscriminately, but ignore certain flies and beetles, and also the eggs and young larvæ of most butterflies.

The second objection, that of accounting for survival during the initial stages of the evolution, has been raised by many critics. It is postulated that if the animal does not resemble part of its environment more or less closely, it cannot exist, and yet in the early stages, when the resemblance was much less exact than it is to-day, its owners survived. Thus we must believe that on the one hand the vision and discrimination of insectivorous animals are so sharp that only perfect disguise can deceive them, and on the other that a general approach

DESCRIPTION OF PLATE IV.

Protective coloration in neotropical Lepidoptera. (All about natural size, except 1 and 2, which are greatly enlarged.)

1, 1 A, Notodontid larva in warning attitude with everted osmeteria (from a coloured sketch). 2, 2 A, Sphingid larva in warning attitude, displaying ventral markings (from a coloured sketch). 3, *Stenoma* sp. 4, Limacodid larva (photograph from life). 5, *Automolis diluta*. 6, *Eucereon metoidesis*. 7, Larva of *Prepona* sp. (photograph from life).

PLATE IV

PROTECTIVE COLORATION IN NEOTROPICAL LEPIDOPTERA

to the model is good enough, and that as long as the subject is moderately obliterative, its enemies will overlook it. To revert yet again to the Membracidæ, ontogeny offers an analogy to phylogeny. The nymphs of this family are cryptically coloured brown or green, and the structural modifications peculiar to the adult do not appear until the last moult. Funkhouser remarks that a few birds have been seen to eat the nymphs to the neglect of the adults in the colony; but he attributes this selection, not so much to the cryptic appearance of the latter, as to distrust of the spines with which the pronotum is furnished. After five months in the tropics, I came to the conclusion that any insect coloured green or brown, and with a moderately irregular outline, stood a fair chance of being overlooked amid the wealth of vegetable forms in the jungle. Membracids blend with their background of leaves, but this end is attained by every modification of shape, round, triangular, or irregular. In one species, *Stegaspis galeata*, the outline of the sexes is different, but anyone who has seen them in natural surroundings will admit that one is as good a representation of a dead leaf-bract as the other (Pl. III, fig. 11, C and D). Poulton remarked that the horn of the thorn-like *Umbonia* is less pronounced in the male, and he suggested that the more exact resemblance of the female might be correlated with greater need for protection. This explanation would be more satisfactory if *Umbonia* lived exclusively on plants whose thorns they imitated closely; but these Membracids live as often as not on plants that are thornless, and escape notice, not by the perfection of their disguise, but by their obliterative coloration (Pl. III, fig. 4).

The third difficulty is the necessary outcome of the postulate of stringent selection; for, logically, the more perfect the imitation becomes, the less strict will be the selection controlling it, because its possessor will stand a better chance of eluding notice. Thus when a colour pattern has been brought up to a certain standard, the selection which hitherto guided it will ease off automatically, and leave that particular character of the organism to vary within survival limits.

These three difficulties, though considerable, are not insurmountable, and the most hopeful path towards the solution of these and other problems of coloration seems to be by the correlation of systematic and bionomical study of a particular character within a natural group. This has been attempted by Dixey (20) for certain butterflies and by Lesne[1] for some African beetles, and as our knowledge extends, it may be possible to work out other groups in the same way. For instance, in the Membracidæ, the line of evolution which culminates in the ant-like *Cyphonia* can be traced back through nearly related forms[2] to species which are non-mimetic but which sometimes have conspicuous (aposematic) colouring. The mimetic *Heteronotus* group contains both black "ant" and yellow "wasp" forms, but the intermediate stages no longer exist, or at least are not known.

Vignon (46) has attempted to trace the development of pattern in neotropical grasshoppers (*Pycnopalpa*), but unfortunately he treats of two species only, and his paper is not illustrated. According to Vignon, these insects are brown and grey in colour, and suggest dead semi-skeletonised leaves. In *P. bicordata*, a pale spot with a rusty border at the base of the elytra simulates the onset of decay, and the pattern is borne out by the femur where the latter crosses the patch. The posterior extension of the hind wing is bordered by a black and apparently purposeless mark like a spot of mould. Evolution has proceeded further in *P. angusticordata*, for the pale patches on the elytra have become larger and transparent, their rusty border has broken up into a number of small spots, and the black mark on the wing has undergone the same transformation. The general effect is now not merely that of a dead leaf, but of a leaf that has been killed by some burrowing larva, which has gnawed the parenchyma and left a trail of frass at either end of its tunnel.

The extension of studies of this kind, and comparison of systematic position with bionomical values, seem to offer the best hope for the solution of problems of coloration. Success

[1] P. Lesne (1914). "Les Longicornes du genre *Phosphorus*." *Nouvelles Archives du Museum d'Histoire Naturelle*, 5e Série, vi.

[2] The genus *Poppæa* is specially interesting since it contains yellow forms which may furnish material for future Smiliine wasp mimics, if indeed they do not already await discovery.

however will be achieved only by much further work both in the field and in the laboratory.

BIBLIOGRAPHY

(1) ANDRÉ, EUGENE (1904). "A Naturalist in the Guianas." London.
(2) BAILEY, IRVING W. (1922). "The Anatomy of certain plants from the Belgian Congo, with special reference to myrmecophytism." *Bull. Amer. Mus. Nat. Hist.* vol. XLV.
(3) BATES, H. W. (1862). "Contributions to an Insect Fauna of the Amazon Valley." *Trans. Linn. Soc. London.*
(4) —— (1879). "A Naturalist on the River Amazons." London.
(5) BEDDARD, FRANK E. (1898). "Structure and Classification of Birds." London.
(6) BEEBE, C. W. (1917). "Tropical Wild Life." New York.
(7) —— (1919). "Birds of the Bartica District." *Zoologica*, No. 8. New York.
(8) —— (1925). "The Ecology of Kartabo." *Zoologica*, No. 7. New York.
(9) —— (1922). "The Edge of the Jungle." London.
(10) BELT, THOMAS (1874). "The Naturalist in Nicaragua." London.
(11) BEQUAERT, J. (1921–22). "Ants in their diverse relations to the plant world." *Bull. Amer. Mus. Nat. Hist.* vol. XLV.
(12) —— (1922). "The Predacious Enemies of Ants." *Bull. Amer. Mus. Nat. Hist.* vol. XLV.
(13) "Biologia Centrali-Americana" (1896). Vol. V, Coleoptera.
(14) —— (1909). Vol. II, Homoptera.
(15) BREHM, A. E. (1912). "Thierleben," Säugethiere, I. Leipzig.
(16) CALVERT, C. P. (1911). "The Habits, Structure and Transformation of the Plant-dwelling larva of *Mecistogaster modestus*." *Ent. News*, vol. XXII, Nos. 9 and 10.
(17) CALVERT, C. P. and P. P. (1917). "A Year of Costa Rican Natural History." New York.
(18) CHAPIN, J. P. (1923). "Ecological Aspects of Bird Distribution in Tropical Africa." *Amer. Nat.* vol. LVII.
(19) CHODAT, R. and CARRISSO, L. (1920). "Une nouvelle théorie de la myrmécophilie." *Arch. sc. phys. nat. Génève*, t. III.
(20) DIXEY, F. A. (1894). "The Phylogeny of the Pierinæ." *Trans. Ent. Soc. London.*
(21) FORBES, H. O. (1885). "A Naturalist's Wanderings in the Eastern Archipelago." London.
(22) FUNKHOUSER, W. D. (1917). "The Biology of the Membracidæ of the Cayuga Lake Basin." *Mem. Cornell Agric. Exp. Station*, vol. II.

(23) Howes, P. G. (1917). "Tropical Wild Life." New York.

(24) Imms, A. D. (1913). "On *Embia major* nov. sp. from the Himalayas." *Trans. Linn. Soc. London*, vol. II.

(25) —— (1920). "On the Structure and Biology of *Archotermopsis*." *Trans. Roy. Soc.* vol. CCIX.

(26) Kaye, W. J. (1906). "Notes on the dominant Müllerian group of Butterflies in the Potaro District." *Trans. Ent. Soc. London*.

(27) Lang, Herbert (1924). "Into the Interior of British Guiana." *Natural History*, vol. XXIV, No. 4.

(28) Lutz, F. E. (1912). "A Membracid and Mimicry." *Journ. New York Ent. Soc.* vol. XX.

(29) Mottram, J. C. (1916). "An Experimental Determination of the Factors which cause patterns to appear conspicuous in Nature." *Proc. Zool. Soc.*

(30) Müller, Fritz (1879). "Ituna and Thyridia: a remarkable case of mimicry in butterflies" (Trans.). *Proc. Ent. Soc. London*.

(31) Picado, C. (1911). "Bromeliacés épiphytes comme milieu biologique." *C. R. Ac. Sci.* t. CLIII, No. 20.

(32) —— (1913). "Les Bromeliacés épiphytes considérées comme milieu biologique." *Bull. Sci. Ent. Fr. et Belgique*, sér. 7, t. XLVII.

(33) Poulton, E. B. (1887). "The Experimental Proof of the Protective Value of Colour and Markings in Insects in reference to their Vertebrate enemies." *Proc. Zool. Soc.*

(34) —— (1890). "The Colours of Animals." Internat. Sci. Ser. London.

(35) —— (1891). "On an interesting Example of protective mimicry, etc." *Proc. Zool. Soc.*

(36) —— (1903). "The Shapes and Colours of the Membracidæ" in "A Monograph of the Membracidæ" by G. B. Buckton. London.

(37) —— (1908). "Essays on Evolution." Oxford.

(38) —— (1913). "A Locustid and a Reduviid Mimic of a Fossorial Aculeate." *Proc. Ent. Soc. London*.

(39) Rodway, James (1895). "In the Guiana Forest." London.

(40) Roubaud, E. (1916). "Recherches biologiques sur les Guêpes solitaires et sociales d'Afrique." *Ann. Sc. Nat. Zool.*

(41) Schimper, A. F. W. (1903). "Plant Geography." Oxford.

(42) Scott, Hugh (1912). "A contribution to the knowledge of the Fauna of Bromeliaceæ." *Ann. Mag. Nat. Hist.* vol. X, 1.

(43) Seitz, K. (1912). "Macrolepidoptera of the World." Pt. V.

(44) Sonntag, C. F. (1922). "The Histology of the Three-toed Sloth." *Trans. Roy. Micros. Soc.*

(45) Swynnerton, C. F. M. (1916). "Experiments on some carnivorous insects, especially the driver ant *Dorylus*, and with butterflies' eggs as prey." *Trans. Ent. Soc. London*.

(46) Vignon, M. P. (1924). "Sur le mimétisme homotypique chez

quelques Sauterelles phanéropterides de l'Amérique tropicale." *C. R. Acad. Sci.*

(47) WATERTON, CHARLES (1903). "Wanderings in South America." London.

(48) WHEELER, W. M. (1913). "Notes on the habits of some Central American Stingless Bees." *Psyche*, No. 20.

(49) —— (1921). "A new case of Parabiosis and the ant gardens of British Guiana." *Ecology*, No. 2.

(50) —— (1921). "A Study of some social beetles in British Guiana, and of their relations to the ant-plant *Tachigalia*." *Zoologica*, vol. III, No. 3. New York.

(51) —— (1922). "Social Life among the Insects." London.

(52) WIDERMUTH, V. L. (1915). "The Three-cornered Alfalfa Hopper." *Journ. Agric. Research*, No. 3.

PART II

THE STEPPE

CHAPTER I

IN the Northern Hemisphere, the desert zone on the one hand, and the sub-arctic conifer forest on the other, enclose great tracts of undulating grass country, which are called "steppes" in the Old World and "prairies" in America. The same kind of open plains are found in the Southern Hemisphere, in Australia, in South Africa, and as the "llanos" and "pampas" of South America.

The largest homogeneous climatic formation of this type is the great steppe[1] of Southern Russia and Western Siberia, which stretches eastwards from the plains of Hungary to the highlands of the Altai. This huge tract of land, subject to local modifications which will be described in their place, is a boundless rolling plain, a sea of herbage, scarcely broken by hills of any size, and transected here and there by great rivers. Those who have not seen it can scarcely form a just idea of the grandeur of the steppe, for in England the widest horizons are interrupted somewhere by mountains, trees, or buildings, and on land at least the immensity of an unbroken skyline is seldom realised. But on the South Russian plains, the houses and townships are few and far between, and they, and such trees as can grow, are hidden in the hollows, so that the eye can pass from one ridge of grassland to another until all melts into the distance. When, as in June, the whole expanse is clothed in flowers of every tint of blue, white, gold, and rose, the sight is beautiful beyond description, and justifies the panegyrics of Tchekov, Aksakoff, Gogol, and other Russian writers on their native scenery[2]. The

---

[1] The word "steppe" commonly means a large tract of unforested and uncultivated, or semi-cultivated country, which has an unfrozen sub-soil, and is covered with herbage for at least the greater part of the year; but many œcologists use it in a more restricted sense, as Rubel (25) has pointed out.

[2] A vivid description of this region is given in a short story, *The Steppe*,

steppe, like the sea in its freedom and space, has power to evoke in receptive and imaginative minds either extreme exaltation or crushing depression. This peculiar power it shares with the

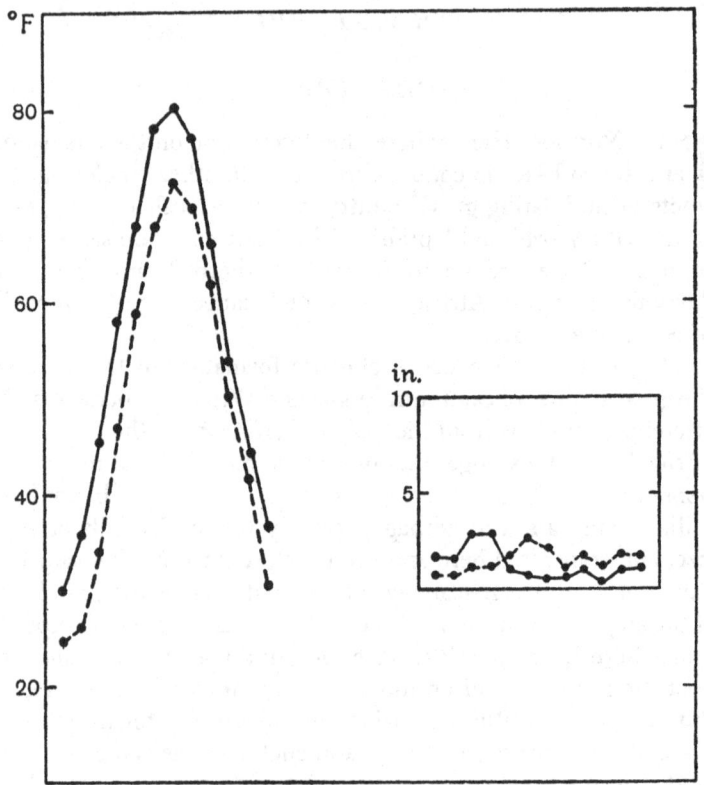

Fig. 6. Average monthly temperature and rainfall of Tashkent and Odessa.
———— Tashkent. – – – – – Odessa.
The comparative warmth of spring, with early maximum rainfall, is characteristic of steppe regions. It is less marked at Odessa, where the climate in the vicinity of the Black Sea is more maritime in type.

desert, its neighbour; and it may be truly said that those movements of social organisation and conceptions of the meaning of human life which have culminated in modern civilisation, all

by Anton Tchekov, published in English in a collection of tales by that writer, called *The Bishop, and other Stories*, translated by Constance Garnett, London, 1919.

arose in steppe or desert surroundings. This is due to the two outstanding characteristics of the physical conditions; namely, the fierce extremes of the climate, and the monotony of the landscape. The first has helped to rear bold, hardy warriors, accustomed to wandering and driven by the menace of drought or famine to stretch out covetous hands towards the nations round them. The second has produced thinkers and dreamers, who turned from the blank plain and sky to inward vision, and thus evolved their own lofty conception of nature and the place of man therein. These two fundamental œcological factors, the variable climate and the monotonous landscape, profoundly influence the character of the steppe fauna also.

The climate of the Eurasian steppe becomes more extreme (continental) the further east we go. The Black Sea region, owing to the proximity of a large body of water, is comparatively equable, although even here the rivers are frozen for about ten weeks in the year, while in summer sunstroke is not uncommon. But at Orsk, in the Urals, the mean winter temperature is about that of the west coast of Novaya Zemlya, while the mean summer temperature is near that of Morocco. Middendorf summed up the steppe climate in four words: "Quecksilber friert; Eier backen." The rainfall is correspondingly variable. Near Odessa, with its maritime climate, there are about sixteen inches annually; but at Irgis the mean is less than half that of Odessa, and the amount becomes progressively less up to the boundary of the Central Asiatic highlands, where a moister mountain climate prevails. For instance, at Tashkent, on the outlying spurs of the Tian-Shan Mountains, the mean annual rainfall is only two or three inches less than at Odessa. The maximum however occurs in March and April instead of in June, and the summer drought is correspondingly prolonged.

It is this question of the annual quantity and distribution of rainfall that really determines whether woodland or grassland shall predominate over the region. For their optimum growth, trees require a moist sub-soil and damp still air in winter. Grass is not affected by dry sub-soil and desiccating winter winds, but it must have frequent precipitations during the vegetative season in spring. Hence it is usually found that in forested country the maximum rainfall is in autumn or winter, while in grasslands it occurs in spring or early summer. The steppe has two drought periods in the year. One is the physical drought of July and August, when actually there is little rain: the other is the physiological drought of winter, when the moisture present is frozen. The only spots where trees can grow are the sheltered valleys round rivers and lakes, where the sub-soil is moist and there is some protection from harsh winds. The open plains and uplands are clothed only with herbaceous plants, which exist through the double drought, either as seeds, or underground as bulbs, tubers, or roots.

The warmth of spring also favours growth of grass. The Eurasian steppe differs from most regions outside the tropics in that the spring is warmer than the autumn (12). In places where snow lies thickly, or where the soil is waterlogged in winter, the spring air temperature curve always lags a little behind the insolation curve, because the cold ground tends to counteract the heat of the sun. But on the steppe the earth is soon swept bare and dry by the wind; and as the atmosphere is not chilled by the soil, the spring temperature readings are relatively high, just as the young grass begins to sprout[1].

In the Black Sea district the onset of spring begins at the end of March, when the snow melts with a suddenness that sends torrents roaring down the slopes and floods the low-lying ground. In April, grass springs up, flowers blossom, birds sing, and insects appear in myriads[2]. Rain falls at intervals, and though the sun is hot there is plenty of moisture. The steppe herbage is not the rich lush growth of our own pastures. It is coarser, scantier, more tussocky, and thickly interspersed with flowering plants, some of which grow three or four feet high. The most characteristic of these are mulleins, mallows, spurges, larkspurs, various pink, yellow, and white Compositæ, *Gypso-*

---

[1] The fact that trees are not found on natural grasslands does not mean that suitable trees cannot grow there, for the Australian *Eucalyptus* and other introduced species flourish on the pampas of South America.

[2] "The spring may seem more potent in tropical lands, but nowhere is it more marvellous than in the steppes, where in its power it stands alone—opposed to summer and autumn and winter." (A. E. Brehm, "From North Pole to Equator," London, 1896.)

phila, Achillea and blue *Eryngium, Scabious, Nigela,* lucerne
and many vetches. May and June are the hey-day of the year.
In July, the flowers fade, the grass scorches brown, and the
steppe appears like a parched stubble field. The watercourses
and lakes shrink to mud-rimmed trickles and pools, or dry up
altogether. Rain falls seldom, and then usually in a torrential
downpour. These rain-storms are often ushered in by a violent
wind, which, rising suddenly out of clouds of threatening black-
ness, sweeps over the plain, enveloping everything in dust. At
its warning the traveller does well to look to the ropes of his
tent, or the fastenings of his dwelling, for after five minutes of
rain the sun-baked earth is a sea of mud of the colour and
consistency of anchovy paste. I recall a motor drive over the
Bessarabian steppe on a hot fine morning, when the dust raised
by the wheels was blown ahead in such clouds by a light
following breeze that it was impossible to see the road surface;
and to avoid breaking springs or axle in the rutways, it was
necessary to tack to and fro on a zigzag course. A thunder-
storm came up, and within ten minutes even chains locked
round the tyres and careful driving did not prevent the car
from skidding completely off the track into a watercourse,
foaming down what before the storm had been a dry hollow
in a millet field. The rain passes as quickly as it comes, but
the downpour is usually so violent that it runs quickly off the
slopes and does little to relieve the drought. In the salt steppe-
desert of Khiva, south of the Aral Sea, more caravans are lost
through rain and hunger than through heat and thirst. Heavy
autumn rains convert the land into a sea of mud in which the
camels are unable to keep their feet; and if the wet weather
continues, the whole company perishes of starvation and ex-
haustion(9). Even in fair weather on the steppe miniature
whirlwinds or "dust devils" are frequent, especially round
villages, and are sometimes strong enough to pick up a haycock,
or empty a threshing floor into the roadway.

The drought lasts throughout August, for though there is
often copious dewfall at night, the heat is intense at noon. The
earth opens in great cracks and vegetation is withered up. In
September the weather is cooler and calmer, and a few flowers

reappear, chiefly shrubby xerophilous plants such as wormwood, goosefoot, thistles, and wild chicory.

The winter is bitterly cold, and a harsh biting wind, called *bora*, sweeps over the country from the east, driving the snow into the hollows and drying up the slopes. The temperature frequently falls below zero, and animal and plant life above ground is at a standstill. Ovid[20] in exile wrote bitterly of the Danube steppe in winter:

> And so great is the power of the North wind awakened, it levels
> Lofty towers with the ground, roofs uplifted bears off.
>
> * * * * *
>
> Ister, with hardening winds, congeals its cærulean waters,
> Under a roof of ice wending its way to the sea.
> Here where ships have sailed, men go on foot; and the billows
> Solid made by the frost, hoof beats of horses indent.
> Over unwonted bridges with water gliding beneath them,
> The Sarmatian steers drag their barbarian carts.

The climate is to some extent modified by local conditions, such as the neighbourhood of open water, forests, or mountains; but, generally speaking, steppe animals have to meet great extremes of temperature, diurnal and annual, with seasonal floods, droughts, and high winds. Their environment is similar in many respects to desert; and in fact the difference between steppe and desert is one only of degree. In the steppe, with its larger or more evenly distributed rainfall, the season of plenty is more prolonged, and perhaps the fruits of the earth when they appear are more abundant; but the time of scarcity is as rigorous in one region as in the other. The steppe can support a larger population than desert, and one that on the whole is more varied and less specialised, for the grasslands are invaded from the north and south by forest and desert-living species respectively. Their advance however is checked by the arid conditions of late summer and autumn on the one hand, and by the cold and damp of winter and spring on the other. Observations by Chernoweth[5] suggest that it is the rate of evaporation rather than the temperature which determines the movements of mammals in moist forests in America; and it is possible that the dryness of the Old World steppe is a barrier to the immigration of forest forms as effective as increased insolation or change of food. Similarly the distribution of vertebrates in the Siberian steppe seems to show that there is a definite limit to the range of desert species into the temperate grasslands.

It cannot be over-emphasised that the resident population of a country is fixed by the maximum that can exist at the least favourable season of the year. To see the Russian steppe in June suggests that its inhabitants must be legion; but as a matter of fact the paucity of species is rather striking, and none is present in overwhelming numbers. Even before man with his herds and husbandry made inroads on the pasturage, it is doubtful if the steppe supported such a large mammalian fauna as the more equable savannas and bushlands of South Africa. To see the steppe in August or in February is to recognise the reason for the scanty population. The climate is so extreme, and the famine so great, that only a certain number can survive. Are these survivals "selected"? In other words, do the weather conditions and deficient food-supply kill off the weaklings and leave the field to the strongest stock? As far as a normal year is concerned, it is reasonable to suppose that this is the case; for when food is scarce the competition to secure it will be keen, and individuals which are crippled, or less resistent to hunger, will die. But this kind of weeding out, though it tends to preserve the individuals who are constitutionally better fitted to endure privation, does nothing to preserve those minute variations which we are led to believe bring about the improvement of the race. The gardener who weeds a border does so by careful hand selection, not with a scythe which cuts down all alike and leaves only the most stunted growth of flowers and weeds. Climatic selection may foster a hardy breed—the Siberian sledge dog can endure hardship which would kill an English collie—but it also eliminates variations, which more particular, though not less strict, selection would have spared. Kropotkin (13) pointed out that the Trans-Baikalian horses are weak and thin at the end of winter, not because there is no food, for even at the worst season there is dry herbage in sufficient quantity to support life, but because it is buried under the snow. The difficulty of obtaining it is however the same for all the horses alike, and the utmost the rigorous winter can do is to kill off those which are most susceptible to cold and hunger. It may be argued that this in itself is advantageous to the race; but besides the normal annual extremes of heat and cold, the abnormal years must be considered, when the stress of drought and famine is much higher. The steppe land is subject to such seasons—the failure of the crops in the Volga basin in the year 1921 was partly due to this cause—and catastrophes on a large scale are not infrequent. Nehring (18), quoting Helmersen, states that in 1827 a great blizzard swept over the region between the Volga and the Urals, wherein the Khirghiz of the Inner Horde lost in two days 280,000 horses, 30,480 oxen and 1,012,000 sheep, besides camels and other livestock. Such disasters, comparable to the effects of earthquakes, volcanic eruptions, and hurricanes on human societies, are probably not selective. They destroy all alike, including those individuals best adapted to normal conditions. Reference should however be made to the observations of Bumpus (3) on one hundred and thirty-six sparrows picked up after a blizzard in America. Seventy-two of the benumbed birds revived, and sixty-four died; and it was found on measurement that the survivors were a little smaller, lighter, and longer in the leg, wing, and breastbone than those which perished. Moreover the extreme variants in either direction were among those that succumbed, so that this particular operation of Natural Selection tended to favour a smaller, lighter race of sparrows, whose range of variation did not diverge very widely from the mean. But apart from the destruction of the fit with the unfit, rigorous conditions do not necessarily make for a healthier race. That which kills the weak may injure the strong, and the survivors may live on with impaired constitutions, which are probably reflected in the number, and perhaps in the physique of the offspring. It is commonly held that the human survivors of famines, sieges, and the like, though they seem to recover from their privations, are "never the same men again," and there is no reason to suppose that animals are an exception to this rule.

The steppe climate varies not only in the year, but from year to year; and it is conceivable that the population of a certain region at a given time may be actually below what the land can support, even in the "hungry" season of a normal year. This uncertainty of the annual cycle cuts both ways; for, given a succession of favourable years, the population, or one of its units, will increase abnormally. The Eurasian steppes have been the cradle of great migrations, which, when they involved man and his ancillary flocks and herds, changed the course of history. It was from the steppe that the barbarians assailed the Roman Empire. In the fifth century, Christendom tottered before the invasion of the Huns, and seven hundred years later, Genghis Khan and his sons of the Golden Horde moved westwards against Europe. These migrations belong to human history: less known, though not less wonderful, are the movements of some of the steppe animals. The irregular migrations of Pallas's sandgrouse were first noticed in the middle of the nineteenth century. This bird breeds in the steppes and deserts of Mongolia, where some individuals seem to be resident throughout the year, while others have a regular migration south in winter into the Gobi desert and China. The sandgrouse are gregarious at all seasons and feed almost exclusively on the seeds of *Chenopodium* and other desert plants. Their affinities lie nearer the pigeons than the grouse. Their plumage is pale sand-coloured, the rectrices are long and tapering, and the legs are feathered. Their flight is so swift that Radde and Prejevalsky agree that they fear no enemies in their breeding haunts.

Before 1859 the sandgrouse was known only from the Asiatic steppes, but in that year a considerable number appeared in Europe, and were recorded not only from Russia, but from Denmark, Holland, Great Britain, and France. In May and June of 1863 there was a larger invasion. A few pairs are said to have nested in the sand dunes of Denmark and Holland, and the birds were reported from places as distant as the Adriatic and Biscay coasts, from Ireland, the Faroes, and Scandinavia. Smaller invasions took place in 1872, and again in 1876; but all previous records were eclipsed in 1888, when a horde of sandgrouse overran Europe, apparently radiating outwards from a centre east of the Volga. The vanguard appeared on the steppes of East Russia in February, and, moving westwards, reached Bohemia early in April, and arrived on the shores of the North Sea by the end of that month. May and June saw the immigrants dispersed over the continent from Archangel to Rome, and in many places the birds attempted to breed. The British Isles received their share of the visitors. Stragglers were recorded as far west as Belmullet and the Outer Hebrides, and there is good evidence that a few pairs laid eggs in Scotland and Yorkshire. The flocks lingered in reduced numbers into the following summer, but by 1890 all had disappeared. There were invasions in the first decade of the present century, but not of the same magnitude as the great migration of 1888. An interesting point in the sporadic western migrations of the sandgrouse is that they have always occurred just before the breeding season, which is at the end of March and April. If the cause of such extensive movements is over-crowding, they might reasonably be expected at the end of the nesting season, at the onset of winter, when the population is most numerous and food is scarce. There is evidence of sporadic autumn migrations also—the birds seem to have appeared in spectacular numbers on the plains round Pekin in 1860—but whether these irregular movements are due to lack of food, over-increase of population, or even to disturbance in the balance of the sexes, is still unknown.

The passenger pigeon used to appear in vast flocks in Canada; but, unlike the sandgrouse, this was part of its regular annual migration, and ceased only with its extinction in the latter part of the last century. Wilson saw a flock estimated to consist of 2,230,000,000 birds; and in the seventeenth century the flocks wrought such damage to the crops round Montreal that the bishop was sent for to exorcise them with holy water (19).

Lowe (14) has recently published an account by an eye-witness of the last great flight of the passenger pigeon near Toronto, about 1870.

By this time the air was black with flock upon flock of pigeons all going eastward....They came on in such numbers that thousands would pass between the discharge of my double barrelled gun and its reloading...millions of pigeons were filling the air, and shadowing the sun like clouds. The roar of their wings resembled low rumbling thunder, and the shooting of scores of guns could be heard for miles, resounding from wood to wood like a small mimic battle.

The flight continued through the succeeding days, though in diminished numbers. For the next few years passages took place on a less prodigious scale, and then ceased altogether. The comparatively rapid extinction of this pigeon is a curious problem; for, as Lowe points out, there is no evidence that it was due to failure of food-supply, nor to epidemic disease, nor to the action of man. The species seems to have run its course and died out after "an expiring blaze of procreation."

The second great œcological factor in the steppe, the uniformity of the landscape, is modified by conditions, local or otherwise. Where the grasslands meet the southern fringe of the forest belt the country is park-like, and as richly wooded, pastured, and watered as the county of Kent. Local formations of the same nature occur elsewhere in river valleys, or where mountain ranges induce a higher rainfall. Thus the southern spurs of the Urals carry a great line of conifer forest into the heart of the Russo-Siberian steppe; and Radde (23) has described the same modification of landscape and climate to the north of the Persian and Afghan highlands, where the great oases, Merv and Achal Tekkes, lie at the foot of the mountains, and are watered by their streams. The southern borders of the steppe, which merge into the Great Palæarctic Desert zone, for eight or nine months of the year are as parched and sterile as any sandy desert of popular conception.

The season plays a part in producing local desert formations. In August the fertile grass steppe round the Black Sea takes on the characters of semi-desert. The grass withers, and its place is taken by thistles, wormwood, and other shrubby xerophilous plants, able to withstand heat and drought. The soil is an important factor. Where it is heavy enough to retain moisture, vegetation is abundant, but sandy ground which is poor and light will support only plants of desert type. Dzinbaltowski (7) found that in Poland steppe formation is largely due to drainage, and that typical grass (*Stipa*) steppe occurred on southern slopes where the water ran off rapidly, especially after the snow melted.

This led to excessive erosion, which increased the dryness and poverty of humus in the soil.

Œcologists are accustomed to distinguish different kinds of steppe formation according to the dominant type of vegetation. From a faunistic point of view the most marked of these are parkland, stipa-steppe, artemisia-steppe and sandy steppe.

Nazarov(17), who studied the distribution of birds in the Khirghiz region, recognised these four formations; and found that, although the area surveyed was comparatively small, each had a characteristic avifauna, especially in respect of eagles and larks.

(a) The parkland or "wood-island" region (région des forêts-îlots) includes the forest-clad slopes of the Urals. Here the grassland is rich, comparatively well watered, and interspersed with woodland, chiefly of the deciduous kind. The fauna is transitional between that of forest and steppe proper. Characteristic mammals are the common squirrel, flying squirrel, the marten, the brown bear, and the suslik. The birds include the imperial eagle, the skylark, the hobby, the sparrowhawk, the great bustard, the woodcock, and many woodland species.

(b) The stipa-steppe is so called from *Stipa pennata* and *S. capillata*, the dominant species of grasses. Here trees disappear except along river valleys. There is abundant vegetation in spring, but drought sets in at the end of June, and animal life is then concentrated round the lakes. The mammals include the jumping rabbit, the marmot, the corsac fox, and the big-nosed saiga antelope; and the birds, the rose-coloured pastor, the crane, the buzzard, M'Queen's bustard, and the sandgrouse. Characteristic of this region are the eastern eagle and the tatar and white-winged larks.

(c) The artemisia-steppe is more arid than the stipa-steppe. Grass grows poorly, and the land is clothed with xerophytes, such as the grey wormwood (*Artemisia*) and a few low thorny bushes. The sandgrouse and M'Queen's bustard range into this formation, which is the headquarters of the eagle, *Aquila glitchi*, and the short-toed and calandra larks. The artemisia-steppe merges gradually into the fourth formation, the sandy steppe, which is identical with dry desert; but even here, in

PLATE V

(*a*)

(*b*)

(*a*) A VALLEY AND (*b*) A MUD RAVINE IN THE BESSARABIAN STEPPE

the most arid spots, Nazarov observed an eagle, *Aquila bifasciata*, and the shorelark, *Otocorys alpestris brandti*. The mammals are characteristic of desert country, such as a jerboa, some gerbilles, the eared hedgehog and the sand hamster (*Cricetus arenarius*). The only exceptions are the common hamster, and the badger, which last, according to Nazarov, is able to exist in these arid surroundings by devouring the hedgehogs.

The observations of Nazarov, though they refer only to a small and peculiar part of the steppe region, may be taken as representative of the changes in vegetation, with the corresponding change in fauna, which occur in this geographic formation as a whole.

It is an œcological paradox that the dry steppes and semi-deserts of Eurasia possess a rich semi-aquatic or lacustrine fauna, centred round the lakes and rivers. The larger bodies of water influence the local climate and flora to such an extent that the character of the surrounding country is often changed. In Bessarabia, the rolling grass steppe comes down sharply to the left bank of the Danube, while on the other side of this arm of the river the rank growth of rushes and willows in the Dobrudja marshes is an entirely different œcological formation. The same thing may be seen elewhere on a smaller scale. In walking over the steppe one frequently comes unexpectedly upon a broad shallow valley, at the bottom of which lies a small marsh or lake, surrounded in summer by a ring of green reed beds, or willow and alder thickets. These steppe lakes have often merely a seasonal existence. In spring they are fed by the snow thaw, which, pouring down from the high ground, cuts deep rugged gullies in the slopes around; but during the summer they may dwindle into a marsh, surrounded by cracked rough mud, and cumbered with the stinking remains of transient aquatic vegetation. These isolated tarns play an important part in the œcology of the steppe, and support a rich and varied insect and bird fauna. Travellers have commented on the abundance of such birds as herons, storks, bitterns, mallard, teal, shoveller, tufted ducks, grebes, rails and crakes, plovers, avocets, stilts, harriers, and many warblers and other small passerines which are found nowhere else in the region. The species are similar to, and often specifically identical with, the original marsh avifauna of Britain, whose dwindling remnant we are at such pains to preserve in our fens and broads. In fact the ancient Briton, who hunted over the Suffolk breck country to the edge of the fens, must have passed through much the same kind of country, inhabited by much the same forms of bird life, as the Scythian who rode across the Dnieper steppes[1].

In late summer these marshes are the only green spots in the surrounding sun-baked waste, and then represent the oases of the grass steppe. In the sandy steppe their œcological position is filled by isolated clumps of shrubby plants. Boettger (1) and Radde (23) remarked that the sand is piled against them in heaps on the windward side, so as to afford protection from sun and dust storms to the gerbilles, susliks and lizards who live under these arid conditions.

CHAPTER II

THE immediate results of the general uniformity of the country are the lack of covert for mammals, and the absence of breeding or resting places for birds. The animals are not only exposed to baking sun and bitter winds, but are also always within view of their enemies. The mammals counter these disabilities in two ways; namely either by burrowing, or else by gregarious habits and fleetness of foot. As might be expected, the dominant forms are herbivorous rodents and ungulates, and the former have chosen the first, and the latter the second way to maintain existence.

The steppe has a long list of burrowing rodents. The most characteristic are the susliks which make great warrens, and often do much damage to crops. Several species of the genus inhabit the Eurasian steppes from Hungary to Eastern Asia, and others live in the plains of North America. Typically the mole-rats, *Spalax* and *Ellobius*, are desert-living forms, but two

---

[1] Marr (16) observes that the heaths of East Anglia offer the nearest approach to steppe conditions found in Britain.

or three species occur in the steppe. They are voles which have become so adapted to underground life that in form they are more like moles than rodents. The body is stout and barrel-shaped, the limbs short and powerful, the tail and external ear almost rudimentary, the eyes small, and in *Spalax typhlus* doubtfully functional. The mole-rats do not hibernate in winter, but dig down to a depth of several feet, and live on bulbs and roots which they are stated to store up in their burrows. *Ellobius talpinus* is said to migrate into the river valleys in time of drought, and to return to the uplands in autumn (2).

The eastern steppes and semi-deserts contain several species of jerboas and gerbilles. The most striking is the Siberian jumping rabbit (*Alactaga jaculus*) or "Pferdespringer." This animal, which is not a true rabbit but one of the mouse tribe, has such powers of leaping that it is said that a horseman can scarcely overtake it. The pikas (Lagomyidæ) belong to the sub-order Duplicidentata, which includes the rabbits and hares. They are stumpy little animals, with short ears and a curious call, whence they take their name of "piping hares." Their headquarters are in the highlands of Central Asia, but one species, *Ochotona pusillus*, ranges as far west as the Volga basin, living socially in burrows like our common rabbit. Other burrowing rodents are the marmot, some hamsters and various voles. According to Radde (23) in Trans-Caspia the hedgehog is also fossorial and lives in burrows eight or nine feet deep.

Shelford (28) cites Thompson Seton's survey of the mammals of Manitoba (27) to contrast the "levels" of prairie and forest respectively. The results, which are given in tabular form on p. 134, show the high proportion of subterranean forms on the open plains as compared to the forest, and the relatively large percentage of gregarious species.

The larger mammals have developed the second kind of protection. The steppe is pre-eminently fitted for a population of social ungulates; and although civilisation has left few on the European steppes, Western Siberia still supports the curious saiga antelope, the goitred antelope, two or three species of gazelles, the maral stag, and, on the spurs of the Central Asiatic mountains, the wild ass. The tarpan, or wild horse of the Khirghiz steppes, is now extinct. This animal, which was allied to Prejevalski's horse of the Gobi Desert, was probably inter-bred with a domesticated strain (15).

		Forest forms %	Steppe forms %
Breeding	Arboreal	26	0
	Ground level	32	30
	Subterranean	42	70
Habit of life	Arboreal	26	0
	Ground level	68	53
	Subterranean	6	47
Social habits	Gregarious	16	47
	Slightly gregarious	26	12
	Non-gregarious	58	41

There are other mammals in the steppe, but they are less characteristic, being adaptable forms from the forest or desert zones. Such are the hare, the wild boar, the roebuck, the pole-cat, some weasels, the badger, the common and corsac foxes, the glutton, the wolf, and the serval.

The habit of life of the birds follows that of the mammals. The dominant species are either able to run or hide swiftly in grass, like the bustards, quail, crane or larks; or they are forms which live on the wing, like swallows, martins, bee-eaters, harriers, hobby-falcons, or buzzards. There are comparatively few perching birds.

The lack of nesting places is an important factor. Many birds breed in holes for shelter and protection from the sun and from their enemies. Such are the sand-martin, roller, bee-eater, hoopoe, and rose-coloured pastor. Even the larks, wheat-ears, and chats build under shelter of stones or clods.

But the larger birds can find neither lofty trees nor crags. The nearest equivalents are the mud ravines which the spring rains carve out of the steppe slopes. These gorges have rugged walls of sun-dried earth, and are frequently not more than fifty feet wide and the same in depth. In spring, they harbour a strange assortment of birds who can find nowhere else to lay their eggs, and apparently live harmoniously together—hawks, sand-martins, rollers, bee-eaters, and jackdaws.

A walk over the Russian steppe at midsummer reveals the limitations of the avifauna. On the high wind-blown ridges, the only birds in sight are a buzzard or two, mere specks overhead, or a mousing kestrel vibrating against the skyline. If the surface is broken ever so little to give vantage-points and shelter, a wheatear will flirt his tail, and a few little dusty larks peck and creep among the clods. But the birds increase in numbers with descent to lower ground, until in the hollows, where a few stunted robinia or apricot trees grow, they become quite numerous. Quails abound wherever the herbage is high enough to afford covert; and their monotonous call, like the flick of a whip lash, resounds from dawn till dusk when nightjars come forth to skim and *churr* over the flowery fields. Hoopoes are common in such places, and the bramble brakes harbour a few warblers. Wherever there is sufficient tree growth, golden orioles and shrikes settle and breed; and serins, goldfinches, and linnets make these coppices their headquarters.

The four characteristic birds of the Russian steppe are the red-legged hobby, the roller, the bee-eater, and the common sand-martin.

The male hobby, with grey plumage contrasted with chestnut, is one of the handsomest of falcons, and flies with wonderful speed and grace. Where a road or railway crosses the steppe, hobbies are always to be seen perched on the attendant telegraph wire, poised for instant pursuit of their prey, which consists chiefly of grasshoppers and other large insects in the grass below. Their neighbours on these occasions are the rollers, handsome and rather sedentary birds about the size of a jay, but conspicuously coloured blue and brown. The roller is clumsy and ungainly on the wing, and usually sits motionless on its perch, from time to time dropping awkwardly down upon an insect. Its success is sometimes baulked by the hobby, which swoops down and carries off the prey from under its bill; but there is no bickering over this, and in a few minutes both birds settle down amicably on the same six feet of telegraph wire.

The bee-eater combines the splendid colour of the roller

with the graceful flight of the falcon. This species is truly gregarious, and small flocks of the slim green, brown and primrose birds are a common sight on the steppe. In Bessarabia grazing herds of horses and cattle are attended by companies of rooks, which turn over the dung in search of beetles, etc. The bee-eaters accompany them, mincing and sidling at a respectful distance until some insect is put up too swift for the clumsy rook to capture. Then the troop of bee-eaters give chase, turning and skimming like swallows. The rooks are rather suspicious of their gay followers; and I have several times seen them drive off a bee-eater who showed too acute an interest in dung-digging operations. Before a storm the bee-eaters become intoxicated with excitement. As the clouds bank up, the birds gather into a flock, and rise almost out of sight in a mazy dance, while the steppe rings with their liquid cries. Usually bee-eaters are rather silent, and this may be said of many of the steppe birds. The explanation probably is that in that great open country, sight supplements hearing, and there is little risk of a bird losing its place in the flock.

The sand-martin is widely distributed over the Palæarctic region, and its abundance in the steppe is not due to particular adaptation, but to the presence of convenient nesting grounds in the banks of rivers and gullies, and to the teeming insect life which is increased by the primitive ideas of sanitation in the human habitations. In summer the villages are infested by flies, bred in the farmyards and open refuse heaps, and flocks of hirundines hawk round the houses. In late summer the sand-martins in the districts round the Danube Delta are an unforgettable sight. One day in 1917 the telegraph wires beside the causeway road across the Pruth marshes to Galatz were laden with an unbroken row of sand-martins, sitting shoulder to shoulder, for seven miles. Others, disturbed by a passing motor car, darted to and fro like a swarm of bees, until the onlooker's senses were bewildered by their cries and rushing wings.

The birds soon take advantage of any artificial work which relieves the uniformity of their surroundings. The hobby, roller, bee-eater, and kestrel have adopted the telegraph pole so whole-heartedly as to make us wonder how they fared in the days before electricity. A shepherd's ruined shieling far out on the open steppe soon attracts a pair of wheatears: a clump of stunted robinias planted by a threshing floor brings flycatchers, orioles and chats; and the roof and eaves of the dwellings are populated by white storks, swallows, and house-martins. During the war military works on the Danube were soon occupied. After the Dobrudja retreat of 1916–17, the Russians dug emergency trenches above the river, and when they were empty the following spring, several pairs of wheatears took possession of them. At the same time, excavations for gun emplacements at the waterside were taken advantage of by a thriving colony of bee-eaters and sand-martins, who entirely ignored the constant firing of an anti-aircraft battery close by. It is not man only who provides artificial covert and protection. Hudson[10] says that on the South American pampas, the vizcacha throws up great trenches round its burrows. These earthworks are used as nesting banks by a species of sand-martin (*Atticora*), and are also inhabited by a burrowing owl, five or six kinds of sand-wasps, a Reduviid bug, and a nocturnal Cicindelid beetle, all of which are seldom found elsewhere in the country. In North America, owls and rattlesnakes are sometimes found in the burrows of the prairie dogs, and this is given as an instance of commensalism in some text-books. But it has been stated[2] that this is a mistake, and that the rodents generally desert their holes when the snake enters. Ingersoll[11] says that they will even unite to shovel sand into the burrow and smother him. The owls usually choose an empty hole to nest in, and when attempts were made to keep the birds and prairie dogs together in captivity, the latter killed the owls. Darwin[6] mentions a bird allied to the oven-bird which generally nests in holes in the ground, but which took so readily to the mud walls of the houses in Bahia Blanca that it became a domestic pest, and riddled the dwellings with its burrows.

In the wilder parts of the Eurasian steppe, the only natural covert is often the reed beds and thickets round the lakes and river banks. The wolves, antelopes, and wild boars alike resort there to bring forth their young; and until recently at all events, in the desolate tracts round Lake Balkash, the tiger still sought the same asylum. In Bessarabia, in September, lakes of only a few acres in extent support an astonishing number of birds, including northern migrants from wild geese to willow warblers, which can find sanctuary nowhere else.

In the arid steppes of Turkestan, the watersheds are chiefly sandy desert for the greater part of the year, while enormous reed beds lie along the courses of the great rivers. The importance of these reedy swamps has been recently emphasised by Uvarov (29) in a remarkable study of the bionomics of locusts. His conclusions are of such biological and economic importance that they may be summarised here.

A locust of non-migratory and non-gregarious habits, *Locusta danica*, is widely distributed over Eurasia, Africa and Australia, between lat. 60° N. and 60° S., except in thick forests, high mountains, and sandy deserts. It has been held to be of no particular economic importance, and its bionomics have not hitherto attracted much attention. Two other locusts, which occur periodically in vast devastating swarms, have long been known from the same regions. They are *Locusta migratorioides* of Africa, and *Locusta migratoria* of south-western Asia. It used to be considered that these three locusts were distinct species; but Uvarov has brought forward strong evidence to show that *migratoria* and *migratorioides* are nothing but phases or forms of *danica*, which appear in vast numbers in certain generations when environmental conditions favour their development. Experimental work has been carried on with the *migratoria* form in Russia and Turkestan, where it is an important pest of agriculture; and research has shown that the breeding grounds are limited to the deltas of the rivers Volga, Ural, Kuma, Terek, Syr-Darya, Amu-Darya, etc., which drain the Black Sea, Caspian, Aral, and Balkash basins. Uvarov describes these river mouths as follows:

These deltas as a rule extend over vast areas irrigated by numberless channels which change their course about every year, some of them forming temporary pools and small lakes. The shores of these channels, and even the beds of the shallower ones, as well as the less elevated portions of land separating the channels from each other, are covered with a dense growth of the gigantic cane or reed-grass (*Phragmites communis*) sometimes 10–15 feet high, which forms almost impenetrable jungles extending over hundreds of square miles. These reed beds however are interrupted, since many of the islands between the channels are more elevated above the level of the water than is suitable for the growth of the reeds, which require a very damp soil. The soil of such islands mostly contains a large proportion of sand, and a peculiar flora of low mesophilous and xerophilous grasses covers them. It is in such localities that *migratoria* lays its eggs as a rule[1], while its larval swarm wanders all over the valley, penetrating the reed beds, and even swimming across the streams. The leaves of the reeds, which contain a very large percentage of silica, represent the most preferred food of the larvæ. The climatic conditions in these reed beds are very peculiar, and differ very much from those prevailing in the adjoining steppes and deserts; since the waters of the river are there spread over a vast surface, the evaporation is very extensive, and the hot air amidst the reeds at midday vividly recalls the tropics.

L. danica is distributed over this region, and it is supposed that under certain conditions it gives rise to the *migratoria* phase. What these conditions are is not certain. Plotnikov[22], in the course of some interesting breeding experiments, claims that, having eliminated the possible effects of food, light, and moisture, nymphs kept in numbers in a cage together developed into the *migratoria* phase, while those kept solitary became *danica*. What agency brought about this remarkable result is not known. In a "swarming" year, *migratoria* appears in great numbers, and moves through the reed beds, the swarms rolling together like a snowball, until the whole horde starts on the march. Again the factors controlling this apparently concerted movement are not known. It is not food, for the swarm will sometimes leave a fertile area for an arid and sterile one. Uvarov thinks that temperature has something to do with it, for unless the thermometer is above 13–15° C. the locusts do not move. In the morning, as soon as the sun gives the optimum warmth, they begin their march and pass forward without feeding, but towards evening, as the temperature drops to 15° C., they stop and begin to eat. So the days pass until the nymphs

---

[1] "The matter is really more complicated than this, since the selection of spots for oviposition is usually closely connected with the more minute characters of the soil and vegetation."

undergo their last moult and emerge with fully formed wings, though for the next day or two they are too soft for flight. At this time the locusts feed voraciously, but one group after another takes wing, and soon the whole horde departs in the air. This flight appears to be always in a definite direction, and although not of very long duration, the velocity is such that great distances may be traversed. Finally the swarm alights and the females oviposit in the ground. But from their eggs arise, not the swarming *migratoria*, but the solitary *danica* phase. This is the more remarkable when we remember that *migratoria* and *danica* not only differ sufficiently in colour, form of legs, elytra, head, thorax, etc., to have been regarded as separate species, but also in behaviour. *Danica* never under any circumstances "swarms," and if a flock of *migratoria* crosses its path it endeavours to avoid them, whereas the ruling instinct of *migratoria* is to crowd together with its fellows. The bionomics of *migratorioides* in Africa are less understood; but there is reason to believe that they are similar to those of *migratoria*, and that its breeding places are in some of the tropical and sub-tropical jungle swamps. Further, there is evidence that some locusts in other parts of the world, *Locustana pardalina* in South Africa, for instance, have likewise a swarming and a solitary form.

It is clear that *migratoria* or *migratorioides* is the distributive phase, and when the range of the species has thus been extended, the position is consolidated by the adaptable *danica* form. If suitable environment happened to be present at the new limit of the range of the species, *migratoria* would again develop in course of time and the cycle would repeat itself. But fortunately for mankind in temperate climates, the optimum conditions for the swarm phase occur nowhere but in these curious local formations of south-western Asia; and although the locust has spread over Europe, the environment forbids any disastrous development. It is worth remarking that up to the eighties of the last century, the Don, Dnieper and Dniester deltas were breeding grounds of *migratoria*; but nowadays, owing to the extension of horse and cattle rearing in these areas, and the consequent destruction of the reed beds, the only per-

manent breeding ground of the migratory phase in the Black
Sea basin is the Danube estuary.

The evolution of these locusts affords grounds for specula-
tion. Uvarov is inclined to think that the *migratorioides* form
—the most constant—is the oldest of the three, and evolved in
Africa, whence it spread to Europe and Asia. Where environmental conditions were favourable it gave rise to the *danica*
phase. But in the south Russo-Siberian area, where the local
environment approached that of its original home, the swarming
gregarious form reappeared, though, because conditions did not
exactly repeat themselves, the structure differed somewhat from
the parent type. At first sight this looks like an instance of
Dollo's Law of the Irreversibility of Evolution[1], but in reality it
is due to another cause. The difference between *migratorioides*
and *migratoria* is one of degree rather than of kind, and, as
Uvarov points out, can be accounted for by regarding the reed
beds of warm-temperate Eurasia as understudies of, rather than
substitutes for, the ancestral swamps of tropical Africa. This
study of the locust illustrates admirably the joint importance
of taxonomy and œcology in bionomical research.

The Eurasian steppe has a large reptile fauna which changes
according to the character of the country. The grass regions
round the Black Sea possess several forms that are found on
dry heaths and meadows in Britain, such as the blindworm,
the common lizard, the viper, and, in the damper valleys, the
grass-snake, which hunts for frogs round the pools. Further
east, the steppes and deserts of the Caspian basin have a more
varied and highly adapted reptile population living in the sand.
The list of snakes includes species of the poisonous *Naja*,
Echis, and *Vipera*, the last of which, according to Zaroudnoi[(30)],
used to cause many human deaths annually, and many non-
poisonous forms, such as the interesting burrowing snakes
Eryx jaculus and *Typhlops*. The lizards of this arid zone
belong to such genera as *Eremias*, *Agama*, *Varanus*, *Gymno-
dactylus*, *Lacerta* and *Phrynocephalus*. There is also a tortoise,
Testudo horsfieldi, which is able to exist on these sandy plains.

[1] An organism never reverts exactly to its original form, even if returned
to the original environment.

Many of these reptiles show characteristic adaptations to life under desert conditions, such as elongation of the body, hard scaly covering, and feet provided with strong claws for burrowing, or with supporting fringes to the digits. *Teratoscincus scincus* is an interesting gecko which, living in dry sand instead of on rocks or trees, has lost its adhesive finger discs, and in their place has developed lateral lamellæ along the toes, which help to support it when running over the loose soil. These adaptations however have arisen in dry desert rather than on grass steppe, and are somewhat outside the scope of the present essay. They have been discussed by Boettger (1), whose work is summarised by Gadow (8). These reptiles have two resting seasons in the year—summer and winter—and in their active existence they are the quarry of buzzards and eagles. From the observations of Zaroudnoi (30) it seems as if many of the adaptations mentioned above, such as thick scales and ability to dig and run swiftly in loose sand, may have evolved as much through strict selection by birds of prey as in response to the physical surroundings.

As might be expected, amphibians are comparatively scarce on the steppe. The species which are most widely spread in South Russia are the edible frog, which is common in swampy places from the Danube to the Caspian, and the green toad (*Bufo viridis*). This pretty toad has a wide range, extending from the Rhine eastwards over Europe and North Africa, to Central Asia and into China by way of the Himalayas, where Stoliczka found it at an altitude of 15,000 feet. In Bessarabia it inhabits the dry ridges of the steppe, often at a considerable distance from water. It is nocturnal, living in holes in the ground by day and only coming out at dusk. According to Gadow (8) it is entirely insectivorous and very prolific. The eggs are laid in water, in two strings, and Heron Roger calculated that one female might lay as many as 10,000. In these habits lies the clue to the success with which the green toad has invaded the steppe. It is in fact an excellent example of an animal which has adopted, rather than has become adapted to, its environment. Structurally it differs little from other toads, and every habit is typical of those of the genus elsewhere, evolved in damp surroundings. It remains a semi-aquatic animal, and there is nothing to suggest that it could survive a steppe summer, and yet it not only survives but multiplies. But it is naturally subterranean and nocturnal, and so has taken kindly to a twelve hours' siesta in the cool ground, protected from heat and enemies. Then it does not eat earthworms, the supply of which is precarious in drought, but prefers insects, which are abundant. Lastly it is very prolific. True it must have water in spring to spawn in; but spring is the wet season when dry watercourses fill up and puddles form. These are chiefly transient, and the mortality from desiccation among the tadpoles must be tremendous; but here and there an oviposition is successful and peoples the surrounding country with toads. Possibly part of the success of this toad lies in its power of absorbing and retaining water, a capacity it shares with other batrachians, such as the common frog. I have captured scores of green toads on the steppe at night, and noticed that the animals when handled would emit enough water to soak a pocket-handkerchief. Dew is probably the source of the supply.

This species suggests certain problems in the coloration of steppe animals. Green or brown tints, in harmony with the general surroundings, prevail to a large extent, but there are some curious exceptions. Buxton (4) has pointed out that brown or grey is by no means the universal colour among sand-desert forms, and he has commented on the frequency of black as an alternative. Moreover, many of the most "protectively" coloured species are nocturnal, and in the dusk their colour can have little survival value. The green toad is a case in point. It is prettily variegated with green and terracotta brown, and both tint and pattern are exactly those which make it least conspicuous on the steppe; but, as the toad never appears in daylight, its colouring can be no protection against enemies. In this case the colour pattern was probably evolved under other than steppe conditions, and is a good example of the danger of reading protective purpose in coloration without a preliminary inquiry into bionomics.

Birds are satisfactory subjects for the study of this kind of problem, because, as far as we know, their palatability does not vary with their colour; whereas in insects there is always the possibility that a conspicuous species is distasteful to its enemies. When two allied species of birds are apparently similar in habits, haunts and foes, and yet one is conspicuously and the other obliteratively coloured, we are tempted to wonder whether after all the plumage pattern has much protective value. This is the case with some of the larks and wheatears on the steppe and neighbouring deserts. The short-toed, calandra, and crested larks are brown and grey, but the tatar lark of the same regions is black and conspicuous. Some wheatears, such as the Isabelline, are sandy buff at least in part; but the male of the pied wheatear, at any rate, is as striking in appearance as a small magpie, and by his lively movements seems even to draw attention to his bright plumage.

The coloration of the insects presents a different set of problems. The steppe is a paradise of Orthoptera. Grasshoppers of all sizes, colours, and powers of flight spring up in all directions from the herbage, from great locusts which whizz away like snipe for a hundred yards or more, to little crickets which can creep only from one leaf to the next. Among the commonest forms are some of the family Œdipodidæ, which are widely distributed in the Mediterranean region also. When at rest they are brown sombre-looking insects, resembling lumps of dried mud, but when they take flight they unfold underwings of misty blue or flamingo pink. The purpose (if any) of this sudden display of colour is not clear. Some naturalists may compare it to the bright patterns along the flanks of certain tropical tree frogs, and suppose that it is designed to dazzle and confuse a pursuing enemy. While camping on the steppe some years ago I had the opportunity of observing the effect of these grasshoppers on what were, from an evolutionary point of view, quite unsophisticated hunters; namely, the sparrows which had recently established themselves round the tents, and the camp mascots, a family of black kittens. In each case, the bird, or the cat as the case might be, seemed quite unable to detect the grasshopper at rest, but as soon as it took wing it was instantly marked. Frequently the insect alighted, and the pursuer was at once at a loss; but when tentative search in the general direction of the flight flushed it again, the attack was renewed, and after one or two attempts of this kind it was usually captured as it rose. Observations of this kind suggest a verdict of "not proven" for the principle of "flash" coloration.

Another remarkable grasshopper is *Acrida turrita*, a sluggish, elongated insect which feeds on patches of a soft pubescent grass on the Russian steppe. In June, while the grass is still fresh and green, the Tryxalids are green, with a silvery bloom shading into purple on the antennæ and margins of the elytra. At this time they are indistinguishable among the blades and purple panicles of the grass. In August, when the herbage is dried and yellow, the grasshoppers still feed in the same spots, but now they are brown and scorched-looking, in the perfect similitude of bits of straw. Mr Uvarov kindly informs me that this colour change is not uncommon in grasshoppers all over the world, even where, as in this instance, only one brood is reared in the year. It has been suggested that the pigment is derived from the food, in the same way that the colour of the egg yolk, and of the legs and bills of white fowls, depends upon vegetable pigments in the diet(21), but this has not yet been proved.

But by no means all the steppe insects are obliteratively coloured, and some of the Hymenoptera, Heteroptera, and Diptera are highly conspicuous. The bug *Pyrrhocoris apterus*, for instance, sometimes appears in such swarms that the low herbage on which it rests seems to be covered with clusters of scarlet flowers; and the sandy bare spots are inhabited by numerous fossorial sand-wasps and apterous ground-wasps (Mutillidæ) which are often of gay or metallic appearance. In these forms, bright colours are often the advertisement for special defences—stings, stink-glands and the like—but other species attacked by birds have no particular protective devices. For instance, the dung-rolling beetles (Scarabæidæ) are common in Bessarabia, where they are taken by rooks and kestrels. The beetles themselves, being black and comparatively bulky, are fairly conspicuous on the sun-baked soil. But even if they were tinted to match it exactly, they would still betray themselves by the movements of their balls of dung, which, trundling erratically in the scanty grass, arrest the eye at a considerable distance, and would render protective colouring of little value.

From the foregoing it will be seen that the steppe fauna has few, if any, elements peculiar to itself, and the same applies to grasslands in other parts of the world. Almost all the species also inhabit either the forest or deserts to the north or south, and seldom show adaptations which are peculiarly due to steppe life. Indeed, in some cases, the forests seem to have driven their denizens out on to the plains, there to thrive without modification of structure. An opossum (*Didelphys azaræ*) is found both on the pampas of South America and in the Brazilian and Andean forests. Its structure is typically arboreal, and yet those of the race which have inhabited the grasslands for ages are indistinguishable from their relatives in the woods, and when trees are provided for them, they climb as nimbly as ever (10). In the pampas also there is a woodpecker (*Colaptes agricola*) which feeds on the ground and breeds in holes in banks and walls. It is apparently unmodified in structure by its mode of life, though since the introduction of trees it is said to have returned to the nesting habits of its ancestors (19). In South Africa another woodpecker (*Geolaptes olivaceus*), which lives in open country, has become so reconciled to boring into walls and banks that it will not occupy trees even when they are planted.

The Eurasian steppe has been inhabited by man from time immemorial, and there is no doubt that its character has changed considerably even in recent years. Agriculture and transport have made great strides; and even in the last century the country described by the older travellers has altered much. In many places, along the borders of the grass zone, forests have been felled, and this has tended to make the climate dryer, thus giving grassland the advantage over woodland. Nazarov (17), writing in 1886, remarked that at Orsk, when a piece of forest had been felled, the dryness of the climate prevented trees growing up again, even if encouraged to do so. Further east, in some parts of the Caspian area, attempts at artificial irrigation have had to be abandoned, for the increased evaporation of the scanty water supply led to drought. It is not always easy to determine whether the trend of a given area is towards de-afforestation or re-afforestation. In places, especially along the valleys of South Russia, there appears to have been a good deal of secondary planting of trees, chiefly near the villages; and here it is possible to distinguish a kind of savanna or bush-country fauna, which is particularly well marked in respect of the bird life. But the extension of agriculture has had the inevitable effect on the indigenous fauna and flora. Alien forms are introduced, and the aboriginal types disappear; and although it seems that, at all events in parts of West Siberia, the steppe area has actually encroached on the forest boundary, only those forms of animal life which can adapt themselves to the new conditions will survive. In a few generations what is left of the so-called "virgin steppe" and its original fauna and flora will be small indeed.

BIBLIOGRAPHY

(1) BOETTGER, O. (1889). "Ueber die Einwirkung von Klima und Boden auf die Tierwelt." *Zool. Garten*, Bd. xx.
(2) BREHM, A. E. (1914). "Thierleben," Säugethiere, 2. Leipzig.
(3) BUMPUS, HERMON C. (1898). "The Elimination of the Unfit as illustrated by the Introduced Sparrow." *Biol. Lectures at Woods' Hole*. Boston.
(4) BUXTON, P. A. (1923). "Animal life in Deserts." London.
(5) CHERNOWETH, E. HOMER (1916). "The Reactions of certain moist forest Mammals to air conditions." *Journ. Ecol.* (abstract), 1917.
(6) DARWIN, CHARLES (1891). "The Voyage of the *Beagle*." London.
(7) DZINBALTOWSKI, P. (1923). *Acta Soc. Bot. Polonaise*, vol. i.
(8) GADOW, H. (1901). "Cambridge Natural History" (Amphibia and Reptiles). London.
(9) HEDIN, SVEN (1914). "From Pole to Pole." London.
(10) HUDSON, W. H. (1903). "A Naturalist in La Plata." London.
(11) INGERSOLL, ERNEST (1906). "The Life of Animals." New York.
(12) KENDREW, W. G. (1922). "The Climates of the Continents." Oxford.
(13) KROPOTKIN, P. (1910). "Mutual Aid: a Factor in Evolution." London.
(14) LOWE, P. R. (1922). "A Reminiscence of the last great flight of the Passenger Pigeon in Canada." *Ibis*.

(15) LYDEKKER, R. (1907). "Guide to the Specimens of the Horse Family, exhibited in the British Museum of Natural History." London.

(16) MARR, J. E. (1911). In a note to Tansley's "Types of British Vegetation."

(17) NAZAROV, P. S. (1886). "Recherches Zoologiques des Steppes de Kirghiz." *Bull. Soc. Imp. Nat. Moscou*, t. LXII.

(18) NEHRING, ALFRED (1890). "Ueber Tundren und Steppen der Jetzt und Vorzeit." Berlin.

(19) NEWTON, ALFRED (1896). "A Dictionary of Birds." London.

(20) OVID. "Tristia," Book III, Elegy X, Longfellow's translation.

(21) PALMER, L. S. and KEMPSTER, H. L. (1919). "Effects of Pigments on Egg Yolks." *Journ. Biol. Chem.* 39.

(22) PLOTNIKOV, V. L. (1924). "Some observations on the variability of *Locusta migratoria* L. in breeding experiments." *Bull. Ent. Res.* vol. XIV, Pt. 3.

(23) RADDE, G. (1886). "Reisenan der Persisch-Russischen Grenze." Leipzig.

(24) —— (1898). "Wissenschaftliche Ergebnisse der Expedition Trans-Kaspien und Nord-Chorassen." Ergänzungheft no. 126 in *Petermann's Mittheilungen*. Gotha.

(25) RUBEL, E. A. (1914). "Heath, Steppe, Macchia and Garigue." *Journ. Ecol.* vol. II.

(26) SEMPLE, E. C. (1911). "Influences of Geographic Environment." London.

(27) SETON, THOMPSON E. (1909). "Life Histories of Northern Mammals." New York.

(28) SHELFORD, V. E. (1915). "Principles and Problems of Ecology as illustrated by Animals." *Journ. Ecol.* vol. III.

(29) UVAROV, B. P. (1921). "Notes on Locusts of Economic Importance with some new data on the periodicity of Locust invasion." *Bull. Ent. Res.* vol. XIV, Pt. 1.

(30) ZAROUDNOI, N. (1885). "Oiseaux de la Contrée Trans-Caspienne." *Bull. Soc. Imp. Nat. Moscou*, t. LXI.

PART III

THE TUNDRA

CHAPTER I

TUNDRA is a word of Finnish origin, meaning an open forestless stretch of country; but it has passed into the Russian language, and is now the recognised name of that huge tract of land which lies across Eurasia, north of the Arctic Circle, and is continued into North America as the "barren grounds." This was the unknown northern waste which Marco Polo wrote of as "a land inaccessible because of its quagmires and ice."

Typical tundra is treeless, though the name is sometimes used for parts of the scanty stunted conifer forest along the coasts of northern Russia and the lower Lena River. The so-called "Big Low Tundra" of the Samoyedes, which lies on either side of the estuary of the Yenisei River, and eastwards over the Taimyr Peninsula, is swampy and low-lying, without rocks, high hills, or deep valleys. Such country is found not only along the coasts of arctic Siberia and Alaska, but in parts of Novaya Zemlya, the North American islands, and Greenland. In Siberia, where here and there it is transected by mountain ranges, or raised into plateaux, it becomes more rocky or rugged, and is then called by the Russian settlers, the "kammenaia (stony) tundra."

Paradoxical as it may seem, the swampy tundra land has much in common with the deserts of the hotter parts of the world. The distinction is that while deserts are *physically* dry, actually with very little water, tundra is *physiologically* dry, and although there is always plenty of water, yet for the greater part of the year it is frozen up as snow and ice, and is inaccessible to living things. The flora and fauna of the tundra for the most part live in a state of physiological drought.

Another point of resemblance is the shortness of the vegetative season of plants. In deserts this depends on the duration and amount of the light yearly rainfall, which calls forth a large number of quick-blooming annuals and bulbous forms. In the tundra it is determined by the short period of comparative solar warmth, which thaws the snow and makes water accessible to the roots. But this part of the parallel ends here, for the tundra plants are all perennials. The chance of ripening seed in that short chilly summer is not to be relied upon for the maintenance of the species; whereas, in the southern deserts, the safest way for the herb to tide over the dry season in a hot country is as a dormant seed or bulb.

Both tundra and desert are subject to high desiccating winds which bury the land under snow in the one case, and under sand in the other. Moreover they havè each their oases. The fertile places of deserts are those where the scanty water can collect, and the sub-soil is sufficiently moist to allow trees to grow. The oases of the tundra are not the spots where most water stands but where there is least, along the rough pebbly slopes of the higher ground bounding the river valleys. Here the snow melts earliest, and being better drained, the soil is less saturated, and can be warmed by the midsummer sun. It is only on these dryer slopes and plateaux, poor and stony though they are, that flowers appear. In short, the oases of deserts are the moistest spots; the oases of tundra are the driest.

Plants are sure indicators of physical conditions. The adaptations of desert plants for growth in dry surroundings are familiar to most people, and the greater part of the tundra vegetation is likewise adapted to flourish in drought, and often in the same way, although the ground may be actually squelching with water. The earth is perennially frozen a foot or two beneath the surface, and is loaded with raw humus acids. This cold sour soil inhibits the absorption of water by the roots, and in addition the stem and leaves are liable to sudden desiccation by harsh winds or increased insolation. The flora of Greenland possesses so many adaptations characteristic of dry surroundings that Warming was led to compare it with that of the Sahara. The root system is often greatly developed, while the aerial parts are stunted and matted into a compact cushion; the stomata are sometimes concealed; the leaves are either succulent, or else leathery, stiff, and needle-shaped, and frequently have a hairy or waxy surface.

The climate of the tundra is one of the bitterest on earth. Along the Yenisei estuary the snow lies all the year round in the deeper hollows and gulches, and everywhere, two or three feet below the surface, the soil is frozen into a solid block of ice. In winter the temperature often drops to − 50° or − 70° F.;

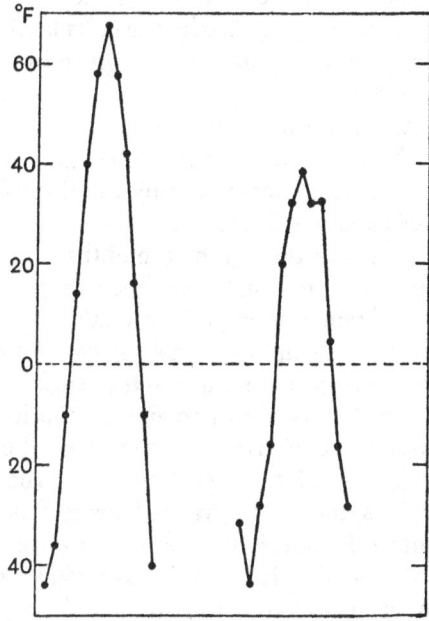

Fig. 7. Average monthly temperature of the Lena delta (right) compared with that of Yakutsk in the forest zone (left). The winters are equally severe, but at Yakutsk the warmer summer allows the growth of trees.

but it is not so much the severity and duration of the cold season which determines the character of the tundra flora and fauna, as the shortness and coolness of the summer. The mean temperature of the coasts of the Arctic Ocean varies as we pass from west to east. The west Siberian tundras are probably swept to some extent by the warmer winds from the Atlantic; and in this connection it is interesting to notice that the coldest part of West Siberia is that which lies under the lee of the

Scandinavian highlands, and is sheltered from the Atlantic winds (19).

On the Yenisei tundras the snow melts at the beginning of June, and for three or four weeks the country is almost impassable. The ice on the rivers, which is often six or seven feet thick, breaks up at the same time, and great blocks, hurtled by the current against the unbroken sheet beyond, are piled into barricades which dam up the waters already swollen by the thaw, and cause extensive floods over the delta land. A vigorous description of this break-up of the ice is given by Seebohm (31). By the end of the month, summer has come. Birds begin to breed and vegetation to grow. From May to July the sun never sets below the horizon, and perpetual daylight reigns. Insect life, such as it is, is at its maximum by the third week in July when the flower oases are in bloom.

July and August are the summer months. At midday the temperature may be quite high—as much as 70° F. There is not a great deal of rain, although occasionally, for a day or two, clouds of driving soaking mist may sweep over the plain, blotting out the horizon and drenching the already water-logged ground. These wet days are often due to the local influence of the great rivers. The principal streams of the tundra, the Petchora, Ob, Yenisei, Khatanga, Lena, etc., run from south to north; and in summer the water, flowing from the hotter regions of Central Siberia, is warmer than the air of the arctic coast. Condensation on a large scale takes place, and results in the formation of masses of cumulus cloud, and in curious mirage effects, comparable to those seen in deserts elsewhere. A hut five miles away looks like a fortress quivering in the air, and its inhabitants appear as tall as trees. When the condensation is excessive, heavy showers and driving mist follow. The end of August sees the last of the fine weather. Gales of wind and rain descend; the last birds depart; day and night are equal in length; the temperature falls rapidly, and snowstorms begin. By the middle of September the land is white once more, and at the end of the month winter settles down. From November to February the sun does not appear above the horizon; and at intervals the tundra is swept by tremendous blizzards—the *purga* of the Siberians—which sometimes last for days together, and pile the snow-drifts up, only to tear them down again and drive them over the plain.

The first impression of the "Big Low Tundra" in June is one of utter desolation. The river banks are flanked either by swamps or by snow-ribbed mudhills. The sandy beaches at the water's edge are strewn with ice blocks, and bleached shattered timber brought down from the forest zone by the floods. The whole landscape, immense though it is, seems repellent, bleak, and formless, without beauty or dignity of any kind. Even the river is so muddy and cumbered with flotsam that it looks more like the titanic scourings of a continent than a noble watercourse. And when the disillusioned traveller lands, and wades through the frozen marshes to the mudhills which bound the valley, the scenery seems still more desolate and inhospitable. We may suppose that the mudhills themselves surprise him a little—they are probably higher and steeper than he thought, and perhaps he begins to realise that in a land so level and so vast, the eye must be schooled gradually to new values of height and distance. Once at the summit, the tundra rolls away before him into the misty horizon, apparently without a single feature to afford a landmark, or to relieve its grey monotony. At this point our traveller may return disgusted to his quarters. But if he perseveres in his exploration, and adventures further, he will revise his first impressions. The fascination of the tundra will seize him, and in after years, though he may recognise the majesty of mountain ranges and the splendour of forests in other regions, he will never admit that they surpass the arctic plains in that mysterious charm which the untrodden ways of the earth possess for those who willingly walk beyond city pavements.

The contours of the country acquire a new importance. Their slopes, razed straight by snow slides, appear steeper and rougher than the real height warrants, and the ascent of the mud bluffs, cut out by the summer thaw, becomes an achievement. Even miles of tramping over moss bogs is stimulating in the fancy that something new and unexpected may lie beyond the skyline, and that the bare and simple landscape conceals romantic possibilities. Sometimes a hill-top view discloses a terraced slope of herbage, as green as that of an English downland, and watered by a stream which rushes between walls of snow-drifts. Or else it is a range of stony dunes, piled up so fantastically that it would scarcely be surprising to see a human sapper at work among them. Little frozen tarns, fringed with gold and green willow bushes in the hollow of the moor, invite exploration; and far away the mirage hangs over a river or lake which scarcely finds a place, much less a name, in any map.

And the weather, which as Stevenson says, is the dramatic element in scenery, is also more varied than first impressions suggested. When the squalls and storms of the thaw have passed, the July days and nights surpass in splendour those elsewhere. At midnight, when the sun dips towards the northern horizon, its light suffuses sky and water with unearthly radiance. The blue flanks of the ice blocks catch the glow, and the icicles drip and fall continually with a musical tinkle and splintering. The splash and gurgle of water racing down the slopes is heard everywhere; and just as the hot blood flows tingling back to a frozen limb, so this warm running water is the life-blood of the tundra, without which it would lie inert and sterile from one year's end to another.

Later, in July, the tundra also sees noons of sabbatical calm when great cumulus clouds drift over the sky, the wind drops, and the horizon is hidden in mirage and blue mist. And in mid-August there are evenings comparable to those of an English autumn, when the year is declining leisurely in peace and sunshine, under sunsets such as we seldom see in the south.

Such halcyon weather is not constant. The rapidity with which the tundra skies can change from blue to grey, even in summer weather, is startling. I have elsewhere(14) described two consecutive days on the tundra in July:

Towards six o'clock in the evening, we descended a long gradual slope from the higher tundra into a river valley. It was raining harder than ever, and the bitter east wind seemed to drive the rain through waterproof and jacket alike, and chilled us all to the bone. We thought gratefully of the *choom* which now lay not far ahead. However, it turned out that we were not to reach it as easily as we expected. The river, though shallow, was wide, and the wind, blowing with the current, drove long tongues of foam down the channel. ...It would have been impossible to find a more forlorn looking party in the whole of Asia. Although it was only six o'clock, and, according to the calendar, still the season of perpetual day, the sky was so lowering that a grey twilight seemed to brood over the wet tundra. Out of the misty mud-hills to the east, the nameless river flowed, no man knew whence, and disappeared among the mudhills to the west, no man knew whither. Little lonely streams, whose sources were visited only by the tundra foxes and the wildfowl, ran down to meet it. We seemed to be the only living things in a land where there was no colour, nor any sound at all except the swish of the wind over the lichens; and as I looked around, I had the fantastic idea that here was a world in the making. It seemed as if in obedience to the enunciation of some great law, the waters of a new chaos had rolled sullenly back, and let the land in all its nakedness appear for the first time. I seemed to see Earth in its beginnings. ...

In the afternoon, the skies cleared, and as I started on a solitary ramble up the valley, I saw the tundra under another guise. Last night, we saw its dour side, its greyness, its loneliness, and seemingly under the scourge of the wind and the rain, its hopelessness. The frame of the land was just as the ice had left it. Its horizons lay in open curves, all angles planed away by the firm hand of the glaciers. Most likely the form of the swamps and rivers had not changed since the mammoth lumbered over the frozen mudhills. But to-day I felt more clearly the promise of the tundra, its huge fertility, its immensity, its strange indefinable magic. Nowhere except in the Alps may be seen such a profusion of flowers—forget-me-nots, lupins, saxifrage, *Pedicularis*, and poppies—and in the hollows the willows were fragrant with bloom. On every hillock, stood a plover in gold-studded livery, playing on his wild pipe, and malingering piteously to lead me from his hidden nursery. Down in the hollow, a pair of godwits whistled to one another in notes like the striking of flint on steel, and red-throated pipits dropped carolling among the flowers. As I walked quickly beside the river bank, little waders ran before me down the sandy spits, too busy to be afraid, and a fine willow-grouse rose with a *whirr* and boomed away over the hill. Beside the ford, the gulls flew to and fro, and stooped at their own purple shadows on the sand-banks. And yesterday there had not been a bird to be seen and all the flowers had hidden their rain-drenched heads. All this transformation had been caused by a little sunshine. It was like a resurrection. The river ran tranquilly, reflecting a clear sky. Yesterday the monotony of its broad flat banks had oppressed us by its drabness and dreariness. But to-day, its very monotony lent it an added charm. It had no history: it served no human purpose. It came out of nowhere, flowed a little way, and then disappeared into nowhere. In some sort the river was typical of the tundra itself. The tundra is a land of the present. It has no past. No history was ever made there, and its people scarcely reckon the flight of years. It has no future. What can you do with a million square miles of lichen and moss which for nine months of the year are frozen fast and deep? The life of the tundra is an eternal present. Thus it was an æon back: thus it will be an æon hence.

Another traveller, Miss Czaplicka(9), wrote of the tundra as follows:

Indeed I forgot my weariness in the beauty of the tundra bathed in the pink glory of the low sun, which rose about one o'clock as we breasted the landward slope of the second range of hills we had crossed in the twilight of evening four hours before. The sun came up as it seemed through an archway of tender light in the slowly flushing east. The dawn had been accompanied by a light drizzle of rain, and as the wind from the river swept this back from us, the level rays traced a perfect rainbow on the gauzy curtain of mist. The dank pools among the sodden mosses and grass, so desolate looking before, were splashed now with all the colours of the post-impressionist's palette. We had not yet reached the crest of the range, and the sky seemed so close that I forgot the world was round, and had a queer feeling of approaching the threshold of infinite space.

And it is not only in the snowless season that the dreary wilderness of the tundra has its hours of transfiguration, even after its quick fading carpet of flowers has vanished before the bleak winds of the short autumn. In winter, too, nerve-shaking as the desolation of the tundra is, when the *purga* shrouds everything in an impenetrable veil of whirling snowdust, there are times when its strange unearthly beauty drowns you in an ecstatic contemplation which quite blots out your aching sense of the terrible cold. There is a rosy light that never was anywhere else by land or sea, that flushes the mountain peaks of the "stony" tundra to an ineffable glory, during the brief twilight days that precede the return of the sun in spring, while the valleys and lowlands are filled with a blue sea of uplifting shadow. And there are the northern lights, which put the stars to shame in one half of the sky, whenever the air is clear, and whether or not the moon is shining.

The œcologist can recognise four formations in the tundra, according to the prevailing type of vegetation. Von Trautvetter(34), from Middendorf's botanical observations, classified them as follows:

PLATE VI

A TUNDRA LANDSCAPE, INCLUDING THE THREE PRINCIPAL FORMATIONS

1. The general level of the drier tundra.

2. The flooded flats.

3. The slopes and declivities.

4. Fertile spots on the sites of old camping grounds and earths of foxes.

From the distribution of breeding birds, I had come independently to similar conclusions on the tundras on the right banks of the Yenisei in 1914, but with this difference, that Von Trautvetter's second group seems really divisible into two, according to the nature of the sub-soil. The first three of these formations are shown in Plate VI. The photograph was taken from the higher tundra overlooking the flats of the Golchika River, a tributary of the Yenisei estuary; and the "flower oases" of the third formation occur along the mudhills on the right of the picture.

1. THE GENERAL LEVEL OF THE DRIER TUNDRA

This region, which comprises vast tracts of higher ground, is bare and dreary in the extreme. The only vegetation is lichen, interspersed with coarse grass, and flowering plants are almost entirely absent. The surface of the land is rough and broken, and the horizon seems boundless. The only relief to the general monotony is an occasional hummock, which covers the den of an arctic fox or affords a resting place to a snowy owl (Von Trautvetter's fourth formation). Here, owing to the enriching of the soil, the herbage is thicker and finer than elsewhere, and makes a welcome patch of green amid the grey lichen. Feilden (13) has commented on these fertile mounds in Greenland; and he remarks what seems to be the case in the Siberian tundras also, that the lemmings resort to them to feed, to the advantage of the foxes, to whose agency they are often due. This high tundra is almost as barren of birds as of flowers; and the curlew-sandpiper, bartailed godwit and the golden plover are the only species which nest there, other than birds of prey.

2. THE FLOODED FLATS

This region includes all the low-lying land along the banks of the rivers which is more or less inundated during the floods of the thaw; but it really consists of two distinct formations, the willow scrub and the moss bog.

The dwarf willow scrub covers acres of ground near the water, wherever the sub-soil is sandy, or at any rate well drained. It is also found in certain places on the slopes of the hills which are somewhat protected from the wind. In such places the willows may grow as much as three feet high, but on the open river banks they are seldom more than eighteen inches, and often less. These puny shrubs represent the forests of the tundra. They are the chief sources of fuel of the wandering Samoyede tribes, and provide the only covert for the nests of the willow-grouse, Temminck's stint, white-fronted goose, and red-throated pipit. These expanses of willow are beautiful in July when the passing wind tosses their foliage into waves of silver and green; but they are still more beautiful at the end of August, when the whole mass turns yellow like ripening corn. This gold of the willow scrub, and an occasional patch of red-leaved whortleberry on the slopes, are the only signs of autumn tints on the tundra vegetation.

The moss bog is found wherever water stands all summer, either along the banks of the rivers, as Von Trautvetter points out, or in depressions of the higher tundra and around the margins of its lakes. Along the estuaries of the great rivers which enter the tundra from the forest belt, driftwood accumulates, brought down by the spring floods. Alluvial soil is washed over this, and so by degrees islands and large tracts of marsh are formed. The surface is quaking and treacherous, but everywhere two feet below there is a solid pavement of ice which never melts. Moss bog is also found in the vast level basins of the higher tundra, into which the water constantly drains from the slopes above. In places it is studded for miles with low hummocks, covered with a spongy mass of sphagnum or lichen. This wet bog is a favourite nesting ground of ducks and wading birds, the little stint, ruff, grey plover, and the phalaropes. In July and August mosquitoes make life almost insupportable in such places. The moss bog is flowerless except for a lousewort (*Pedicularis*) and a few sedges. In some places near the rivers, where the soil is sandy and drier, these sedges appear in some abundance at the end of August. In fact, on a calm evening, when the water of oil-like placidity reflects a clear sky and the sunset tinges this coarse herbage with gold, one may go so far as to compare these sedgy spots with the lush water meadows of our own country, an illusion which is strengthened by the young broods of stints which gather here to feed, and skim from spit to spit with outspread wings just flicking the water, as our barn swallows do when hawking for food at sunset.

3. THE SLOPES AND DECLIVITIES: THE FLOWER OASES

These are found on the south side of the stony hill tops, frequently above the river valleys. The naturalist, rambling along such a ridge in July, comes here and there upon little patches blazing with flowers; for here the snow melts earliest and the pebbly soil is warmed, while there is none of the wet sour humus which partly inhibits plant growth on the level ground. The plants grow in mats and cushions a few inches high, thickly sown with blossoms, white, red, blue, and yellow. There may be seen *Saxifraga oppositifolia, Armeria arctica, Papaver nudicaule, Ranunculus pygmæus, Oxytropis sordida, Myosotis alpestris, Erigeron uniflorum*, and species of *Melandryum, Cochlearia, Draba, Stellaria*, etc. Except for the lousewort in the bogs, and for a coarse butterburr by the river, the flowers of the Yenisei tundras are almost entirely confined to these rocky hills where the insolation is most intense. But here and there, in exceptionally favourable spots, flower patches occur on lower ground, where the aspect is sheltered and sunny, and a trickle of water descends from the hill above. In such a place, in August, I found a bed, small enough to be covered by a tablecloth, of the blue Jacob's ladder (*Polemonium*) and the beautiful orange globe flower (*Trollius*) which is such an ornament of the Siberian meadows further south. These plants were from one and a half to two feet high. Middendorf has commented on the vigorous growth made in these favoured spots by such plants as *Delphinium*, *Sieversia*, and *Sisymbrium*; and records that by Lake Taimyr he found *Senecio palustris* growing abundantly with inflorescences four inches in diameter. But nevertheless, these exceptional records are in themselves an object lesson in the severity of the tundra climate, and the struggle of the plants to maintain existence. For every seed dropped broadcast by bird or wind, which had fallen as these had done where growth was possible, millions must have fallen and perished on the bare lichens of the uplands.

In such a climate, and in such surroundings, the animals of the tundra live and multiply; and this time-honoured phrase is not out of place, for one of the first impressions of summer on the arctic plains is the number of living things that this seeming wilderness can support. Only a closer inspection shows that though individuals are relatively abundant, species are few, and are nearly all representatives, or at any rate allies, of forms which occur further south. For instance, the tundra immediately round the Yenisei supports thirty-eight species of birds. Of these, fifteen species (or sub-species) breed in Britain; six more nest in Norway, as far south as lat. 60° N.; and only sixteen belong exclusively to the high north.

Moreover the most abundant forms of life in the arctic, those which can best support themselves under the rigorous conditions, are all highly organised beings on the top twigs of their respective phylogenetic trees. And yet, with a few exceptions, they are not particularly specialised. They can exist as well, and multiply in greater numbers in the south. They appear to have been crowded out from more genial climates, to struggle for existence in the frozen swamps of the tundra. The term "Struggle for Existence" raises a question about which there are frequently misconceptions. There is a prevailing idea that the struggle, in the Darwinian sense, must be less intense in the tundra than elsewhere, partly because there are fewer competing forms, and partly because so many writers have emphasised the fierce struggle in the exuberant life of the tropics. This is of course a fallacy. Every habitable piece of the earth's surface over an average of years supports a mean number of living beings. The numbers will be higher in a mangrove swamp on the Amazons than on a beach in Novaya Zemlya, but the struggle to survive will be quite as severe in the latter region. In fact, as Darwin originally pointed out, if we believe in Natural Selection at all, we are driven to the logical conclusion that intra-specific must be even more intense than inter-specific competition, because the needs of the contending organisms will be similar.

Suppose that there are two small islands, just alike, and each bearing enough grass to maintain fifty ruminants. Seventy antelopes are placed on one island, and thirty on the other; and in this latter island let us suppose that three or four lions are introduced. The thirty antelopes on island number two will spend a good deal of their time in running away from the lions, but they will have plenty of grass to eat. As for the seventy antelopes on island number one, if they could choose, they would soon be asking to change places, and would be quite ready to put up with the lions, for the sake of a double ration of grass. The struggle for existence does not mean only the necessity to avoid enemies. The factor which counts for even more is the competition for a limited food-supply. Hence in the arctic, where the available food is scanty and there are a number of individuals belonging to relatively few species, the competition between these individuals will be great. Of course the number of beings in a given area will not be as large as elsewhere. Feilden (13) noticed that in Grinnell Land, although the polar hares attained their full size and weight, they were much more sparsely distributed than in regions further south; and he remarked that the probable consequences of the killing of the hares for food by the "Alert" expedition would be that the population would be on the verge of extinction for years afterwards. It is a recognised economic fact in the north that each domesticated reindeer requires 4·44 square miles of tundra grazing per annum; and in the Lapp countries, international commissions have been necessary to determine the limits of the reindeer grounds of the different nationalities.

There is evidence that something of this kind happened in the North Sea during the European War. For some years, much of the North Sea was closed to trawlers, and consequently, on the analogy of arable land, it "went fallow." The fish were not caught, and developed as they pleased. After the War, when trawling began again and figures were available, it was found that the average catches of plaice were larger, and also that the average size of the fish was greater. This looks like a contradiction of the bionomical principle laid down above, for here were bigger fish and more of them than when man was able to interfere. But it was also found that the average age was greater; and that where before the War a fish of a certain centimetre length was three years old, after the War a fish of the same size would be four years old. That is, though the fish grew up and grew larger, they took longer about it because there were so many more mouths to be fed than when man was a factor to be reckoned with, and removed the surplus population (4).

CHAPTER II

THE mammals of the tundra are of particular interest because they have no special protection against the winter cold. Birds are mobile: they can emigrate to a more favourable climate. Insects, at least the great majority of the tundra insects, pass through a resting stage which enables them to survive adverse conditions. The mammals can do either only to a limited extent. It is true that the deer move southwards towards the forest, and probably the wolf follows them; but at best this is but a small amelioration of their condition. The smaller forms must remain in their summer haunts; and, in view of this, it is remarkable that hibernation is unknown among them, although in the forest zone it is common and apparently offers an excellent means of tiding over the cold season.

The nature of hibernation is discussed in Part IV of this book, and it will suffice to say here that perhaps one reason why hibernation is rare on the tundra is that the cold short summer does not allow of the necessary accumulation of food reserves in the body. Tundra animals never have a good balance at the nutrition bank, but, even in the warm season, live right up to their credit, with nothing to put by for a rainy day. A second reason, paradoxical as it seems, is that the great snowfall partly obviates the need for hibernation. Snow is a very bad conductor of heat, and it is a well-known fact that the temperature below the drifts is higher than on the surface. According to some observations made at Petrograd after a fall of twenty inches of fine dry snow, the temperature on the surface of the drift was − 39° F., while that of the soil below was 27° F.(19). It is really because the cold and the snowfall on the tundra are so great that the animals are able to survive at all. On the steppes further south, where the snowfall is less, and the ground is often swept bare by biting winds, a winter sleep is much more usual. The small animals retire into deep holes in the ground below the nip of the frost, and the annual plants hibernate as seeds. Collett has observed that in Norway the lemmings on the mountains do not hibernate, but remain active under the snow. If, however, they are brought down to the valleys and exposed to the frost, they die rapidly. It has also been shown that juniper bushes when covered with snow can survive very severe winters, but any parts exposed above the drifts are blighted.

The most striking mammals of the tundra are the reindeer, the lemming, the arctic fox, the arctic hare, the wolf, the ermine, and, in Greenland and North America, the musk ox, though this last is perhaps more a mountain animal. The Siberian tundras also possess a few species which are really forest forms, such as the glutton, the brown bear, the common fox, and some voles.

Comparison has been made already between the tundra and the desert, and the analogy may be pursued further by comparing the reindeer with the camel. Both are able to live under conditions impossible to most other animals; both have been domesticated from time immemorial, and are still the surest, if not the only form of transport over snow or sand. Without them the tundra and the Sahara alike would probably be still without human occupation. Of the two, the reindeer is perhaps the more remarkable because he does not depart much from the type of his race in more equable climates, and he has adjusted himself with very little modification to arctic conditions.

In summer the wild reindeer ranges up to the shores of the Arctic Ocean, frequenting chiefly the high ground which is comparatively free from mosquitoes. The fawns are dropped in May, and the Samoyedes of the Yenisei say that at this time there is a sort of "Truce of God" between the deer and their enemies the wolves. Both animals resort to the same parts of the tundra to breed, and for a short time they live there in harmony. The bionomical explanation is, not that the wolves have sentimental scruples as the natives suppose, but that in summer lemmings afford them abundant alternative food.

The pasture of both wild and domesticated reindeer consists largely of the lichen *Cladonia rangiferina*. The Siberian reindeer tribes never feed their deer even in winter, but turn them out to fend for themselves. As is well known, the reindeer differs from other Cervidæ in that both sexes have antlers. These are very characteristic, as one or other of the brow tines, sometimes both, are expanded and broadly spatulate. These "snowshovels" enable the deer to rake the drifts aside to obtain food. Reindeer are found of all tints from dark grey to white; and the latter is popular for domesticated deer because it is conspicuous at a distance and allows them to be herded readily. In summer the reindeer, both wild and tame, are in poor condition. The antlers are in velvet, and the long winter coat drops from their flanks in clotted masses. They are sluggish and languid, and devote themselves to feeding and resting to build up strength for the coming winter. The natives of Siberia are loth to harness their deer in summer, and never drive them far in the day. But even in summer snow has an attraction for the deer, and a herd will return from pasture to an unmelted drift to rest and sleep. At the beginning of winter, on the other hand, the deer are in good condition, with fine antlers and coats, and are then capable of travelling a hundred versts at a stretch. The hoofs of the reindeer are somewhat splayed, affording support to the animals on snow or soft tundra moss. Professor Stanley Gardiner tells me that in the peat bogs of western Ireland the hoofs of the donkeys which carry the turf are often allowed to grow long and upturned for the same reason.

The lemming is circumpolar, and is represented by different species in Europe, Asia, and North America. It is the staple food of the arctic birds and beasts of prey, and its disappearance would probably be followed by that of half the vertebrate fauna of the tundra.

The lemming remains active all winter under the snow, and, according to Manniche (23), sometimes even breeds there. At this season its chief enemy is the ermine, which follows it underground. It finds plenty of food in the buried herbage, and as the snow provides warmth and shelter, winter is probably the most comfortable season from the lemming's point of view. Summer life is another matter. When the snow has melted, the lemming can make only shallow runways along the surface, for the frozen soil forbids deep burrowing. Here he is exposed to the attacks of many enemies, wolves, foxes, falcons, buzzards, owls, and skuas. Even the herbivorous reindeer devours him on occasion; though in this instance he should probably be regarded as a tonic rather than a food, to satisfy the deer's craving for blood salts. The distribution of foxes and birds of prey depends largely on the presence of lemmings. The years 1912 and 1913 were remarkable for the numbers of these rodents on the tundras round the Yenisei estuary; and also, according to the inhabitants, for the numbers of skuas and falcons which visited the region. But, when I arrived in the district in 1914, the lemmings had disappeared, although the tundra under the melting snow was covered with their nests and droppings of the previous year, and birds of prey and foxes were correspondingly scarce.

The lemming fluctuates greatly in numbers, and a year of increase frequently leads to emigration on a large scale. These movements have been studied by Collett (6) in Norway. In certain years the number of offspring in a litter is above the average, and the young themselves grow up and begin to breed sooner than usual. The population is at its maximum at the end of the summer when food begins to fail. The animals on the outskirts of the congested area forage further afield: they are followed by other emigrants crowded out from the famine region, and when the pressure from behind is excessive, a great movement may take place. It has been stated that most of these migrants are young males. These "increase" years and "hunger marches" are not peculiar to the lemming, but are known to occur in other species; and the spectacular nature of the lemming migrations is partly due to the nature of the country. The Norwegian mountain lemming inhabits the high fells on the slopes and plateaux above forest growth, and each area is cut off from the next by mountain peaks or steep valleys. Hence the tide of migration is restricted in outlet, and crowds down to the lowlands by certain passes in swarms large enough to attract attention. The smaller wood lemming also increases and migrates in the forests, but the trek is less noticeable because it is radial and the animals quickly arrive in new feeding grounds.

The increase of the mountain lemming does not take place all over the country at the same time, but even local fluctuations may involve areas of considerable extent. It is difficult to say how far the migration may extend, because the uplands of Norway are well provided with lemming grounds; but in 1862, the hordes which invaded Christiania must have travelled at least ninety kilometres. Collett says: "I saw individuals running up the high granite stairs of the vestibule of the University of Christiania as if it were their intention to exhibit themselves in the museum." He also quotes a thirteenth century Norse version of the Book of Exodus, in which the word "locust" in the account of the Plagues of Egypt is translated as "lemming." Manniche (23) found that the Greenland lemming migrated over the frozen ocean to the mainland from islands fifty kilometres distant. Rae (30) writes:

When travelling in June 1851, southwards from the Arctic coast along the west bank of Coppermine River, and north of the Arctic Circle, we met with thousands of Lemmings speeding northwards; and as the ice on some of the smaller streams had broken up, it was amusing to see these little creatures running backwards and forwards along the banks looking for a smooth place with slow current at which to swim across. Having found this, they at once jumped in, swam very fast, and on reaching the other side gave themselves a good shake as a dog would, and continued their journey as if nothing had happened.

At these times the lemmings seem to be guided by a remarkable perversion of the social instinct. Their movements take place chiefly at night, and as a rule the animals proceed singly on their way; but where one lemming has passed, the rest follow, along the rutways and gutters of the roads, down steep slopes, and even over the ends of piers and jetties, where destruction awaits them below. No definite path is chosen. The swarm travels erratically, like a shoal of fish or a flock of birds which follow the accidental movements of the foremost individuals of the band; but of course the general direction is determined partly by the nature of the ground traversed. In some districts they appear in such numbers that crops are spoiled, and barns and even houses are rendered uninhabitable by the accumulation of dead bodies under the floors. Where food is found the march is partly checked, but the lemming cannot survive in the lowlands; and although a few individuals exist through the winter, and have been known even to breed in the plains, cold, disease, and numerous enemies all contribute to exterminate the horde. Water is no obstacle, though there is no truth in the story, repeated even in some recent text-books, that the lemmings march in an army to the coast and commit suicide in the sea. This legend of self-destruction probably arose out of incidents such as that related by Collett in 1868, when a steamer in Trondjem Fiord passed for fifteen minutes through a pack of swimming lemmings. Migrating hares and rabbits are known to swim rivers; and in 1867 a swarm of squirrels is reported to have invaded Tapilsk in the Urals, swimming the Tchussoveia River, and climbing up the oars of boats which crossed its track. Collett says that the lemming is able to cross an arm of the sea more than two miles in width; and in Norway, where the coast is indented by long narrow inlets, it is natural that a migrating army should sometimes try to cross a fiord as if it were a river and be swept away by the tide.

The arctic hare occurs on the drier parts of the tundras and arctic mountains of both Old and New Worlds. In Greenland, according to Manniche (23), it spends a good deal of the cold season under the snow, and feeds on the roots of dwarf willows. It does not hibernate in the strict sense of the word, although it sometimes lies inert for days together.

The arctic fox is circumpolar and is found both on mountainous and on low-lying ground. It lives well in summer on lemmings, and birds and their eggs, and according to Feilden (13) has been known to lay up stores of food for the winter. He found several *caches* in Greenland, one of which contained fifty lemmings, another, half a hare, and the third included the wings of some young brent geese which had been killed the previous autumn and remained all winter in natural cold storage.

But in spite of this there is no doubt that the large carnivores fare most miserably in the arctic winter and drag out a bare existence. Manniche made notes of all the mammals killed in winter during the Danish Expedition to north-east Greenland, and against each we find these kind of entries:

5. ii. o6. Two bears, very emaciated.
10. iii. o6. Young bear, thin, stomach empty.
21. iii. o7. She-bear, very thin.
The chief part of the foxes caught in the winter time were very emaciated.
The emaciated, almost skeleton-like bodies of the animals killed in January, February and March furnished proofs of the extremely miserable conditions to which the wolf is subjected during the long hard winter.

Manniche expressly denies that the Greenland wolf obtains either hares or foxes as winter food; and he supposes the species to be on the verge of extinction owing to the disappearance of the reindeer from that country. In fact it seems as if carnivores in the polar winter suffer the extremity of famine.

Birds are the ornaments of the tundra. If all the mammals, and probably most of the invertebrates were eliminated, their absence would be scarcely noticed; but if the birds were taken away the land in summer would be desolate indeed. They preponderate over other forms of life because they are the most mobile of animals. All species are migratory, and not even the hardiest are able to winter in the north.

On the Yenisei tundras the advance guard of birds appears about the beginning of June, but the main body does not arrive until later in the month. We are still in the dark as to how they travel, but there is some indication that they often follow the courses of the great rivers. The whole business of these birds, for which some of them have travelled as much as three or four thousand miles, is to rear a single brood of young. The time at their disposal is very short. The snow does not disappear until the beginning of July and the winter storms arise at the end of August, so there are at most eight or ten weeks in which to pair, build the nest, hatch out the eggs, and rear the chicks. Some species mate before, or during the journey north, and so save time on arrival. On the Yenisei the black-throated and red-throated divers migrate down the river in spring in couples; and Buturlin ((3) cited) observed on the Kolyma delta that most of the roseate gulls were already paired on arrival.

In spite of their tremendous journey most birds are very fat at the onset of the breeding season. At the beginning of July there is little or no insect food available; and during the first week, which is a period of great excitement, and of constant courtship, rivalry, and fighting for territory, the bird lives almost entirely on its own fat. Later on in the month, when the pools in the marshes have thawed, the staple food of all the insectivorous forms is mosquitoes. It is not too much to say that at least fifty per cent. of the birds of this region, that is, all the Passeres and Limicolæ, are dependent for their summer existence and well-being on Culicidæ. The mosquitoes appear in the first warm days of July, and for about a fortnight life on the tundra is a torment, not only to man, alien and aboriginal, but to his domesticated animals, the dog and the reindeer. Indeed the mosquito directs the annual movements of the Samoyede, and orders his customs and family life; for in summer when the antlers of the sledge deer are in velvet, the animals are so irritated by the insects that it is impossible to keep them in the marshes, and they are taken to pasture on the high ground inland where the plague is less felt. But these swarms of mosquitoes make bird life possible, for they abound everywhere, and can be picked up like manna, not only by the smaller waders and passerines, but also, according to Manniche (23), by larger forms such as the long-tailed duck, king-eider, and ivory gull. Patterson (28) relates that a curlew-sandpiper, kept captive in this country, was very adroit at catching flies, doubtless after much practice on his native tundra. Some arctic birds have rather a wide range of diet. Stints and plovers, if needs be, will eat willow buds: the arctic tern, which in the south takes only live fish, hawks round the fishing stations of the Yenisei for garbage: the skua, and, so it is said, the golden plover, will eat crowberries. The knot, according to Manniche, consumes a good deal of vegetable food.

More than sixty per cent. of the tundra birds have young which are able to run and peck soon after hatching, and fledge quickly. The wing quills of willow-grouse develop so rapidly that the chicks are able to flutter from the ground when only a few days old. The period of dependence on the parents is thus shortened, and the brood is more mobile in search of food. Moreover, in a large proportion of tundra birds, both parents care for the young, and this no doubt assists towards the speedy development of the brood[1]. The tendency to flock together as soon as the chicks can run also ensures additional protection from enemies. This social habit is very marked in the plovers and stints. The nests are usually some distance apart, but the different broods congregate in parties very early, and as the mosquito season wanes, the birds come down from the tundra swamps to the river flats in search of small crustacea, mollusca, and other littoral life. Some of these flocks are composed of one species only. The phalaropes and Temminck's stints seldom

---

[1] It is hardly necessary to observe here that this joint care for the young is not peculiar to the tundra species, but is found in various groups of birds all over the world, especially in passerines and waders. But is worth remarking in this connection, if only as a reminder that the environment belongs to the animal as much as the animal belongs to the environment. Each species inherits from its ancestors, together with its bodily organisation, those complicated reactions to its surroundings, which in their manifestations we call "behaviour"; and in virtue of these it is able, not only to accommodate itself to "habitations enforced" but can enter new habitations. It is inherent in plovers and snipes for both parents to rear the brood, and this, among other things, has helped to make life in the high north possible for them.

associate with any but their own kind. Others, the grey and golden plovers, curlew-sandpipers, little stints, and dunlins assemble indiscriminately. If the feeding ground of such a mixed flock is invaded by an enemy, the old birds show the greatest anxiety. Sometimes a patch of ground only an acre or two in extent is occupied by two or three families of plovers and as many curlew-sandpipers; and birds of both species will stand side by side on a tussock to abuse the intruder. Often all the parents unite in a compact flock, and dash round the enemy until he withdraws.

Besides the protection of numbers, this sociable habit seems to gratify the gregarious instinct that is implanted deeply in most animals. Aristotle remarked that only gods and wild beasts love solitude. This may be true of Olympus, but it does not hold for tundra, steppe and forest. Even those creatures which are commonly supposed to live solitary, such as some birds and beasts of prey, tend to come together in societies wherever they have not been harried into scarcity by man. It is sometimes assumed that a certain species is common because it is gregarious; whereas it would probably be nearer the truth to say that it is gregarious because it is common. On the tundra also are found several instances of that alliance or co-operation between different species or individuals which Kropotkin discussed in his book, *Mutual Aid: a Factor in Evolution*.

The Samoyedes say that the king-eider is a "chief," and that when it is attacked by a buzzard or a fox the other ducks unite to protect it; which is only another way of saying that ducks combine against a common enemy.

The turnstone, which breeds on the arctic and sub-arctic shores of both Old and New Worlds, has earned its name from its habit of turning over stones and seaweed to find its food; and it is often accompanied by purple sandpipers who profit by its labours. When a stone or bunch of wrack is too heavy for one bird, two or three more come to its assistance. Stone-turning is the outstanding habit of this species. Patten [27] relates that two or three turnstones were kept in an aviary with some Californian quails. The nestlings of the latter, which are coloured like pebbles, had no peace, for whenever a turnstone came by, it tumbled the unfortunate chick over to see what lay underneath, and as the result, in a day or two the little quails died of exhaustion.

When Popham (29) found the red-breasted goose breeding on the Yenisei, each nest was placed under a bluff occupied by a pair of buzzards or falcons, which assured the geese some protection from marauding foxes. This habit of the geese is well known to the Siberian settlers. In Spitzbergen, the bernicle goose tries to escape the foxes by nesting on crags, and the goslings descend what is virtually a precipice to reach water (18).

Manniche (23) has given an interesting account of the breeding place of the bernicle goose in north-east Greenland.

Almost all the geese used to leave the marsh at certain times and disappeared southward towards the high middle part of Trekroner. I set out in this direction, thinking that a larger lake was lying near the mountain, and that the geese retired to this after their meal. I really found a pair of larger fresh-water basins and saw in these a few geese, which being frightened, flew further towards the mountains. Having come within a distance of one kilometre from Trekroner, I solved the riddle. The bernicles were swarming to and fro along the mountain wall like bees at their hive, and I heard a continuous humming, sounding like distant talk....While some of the geese would constantly fly along the rocky wall and sometimes mounted so high in the air that they disappeared on the other side of the rocks, the majority of the birds were sitting in couples upon the shelves of the rocky wall, some of which seemed too narrow to give room for the two birds, much less for a nest. It was only on the steep and absolutely naked middle part of the mountain wall that the geese had their quarters, and in no place lower than some 200 metres from the base of the cliff....On the rocky wall in the middle of the goose colony, a couple of gyr-falcons had their nest.....I was surprised to see that the geese were sitting in couples on the projections close to the falcons. When I, by means of a pair of rifle bullets...caused the breeding falcon to fly out of her nest, she and the male circled round the mountain in company with the geese for a long while. From a dizzy height the falcon at last, as swift as an arrow, shot down to the nest, and was soon followed by the bernicles, who again confidently took their seat close by.

The arctic tern breeds in colonies and boldly drives off an enemy many times larger than itself. When Buturlin found the roseate gull in its summer quarters on the Kolyma delta, the nests were placed in "little colonies of from two or three to

ten or fifteen pairs, in company with the common black-capped
tern of the delta, which however in nearly every case exceeds
it in numbers " ((3) cited).

It is well known that in temperate climates most birds begin
to moult their summer dress as soon as their broods are fledged.
But in the arctic summer this process must be accelerated if
the new plumage is to be ready in time for the autumn migration at the end of August. In this country the feathers are shed
gradually and the bird is seldom incapacitated from flight; but
on the tundra, as soon as the young are hatched, the old birds
begin to moult at such a rate that they present a most miserable
appearance at the end of July. Lapland buntings, snow-buntings,
and shore-larks lurk in the herbage, and are almost incapable
of flight.

In some cases, as with the willow-grouse, the moult accompanies the nesting season. In fact the birds spend the summer
in successive moults. In the high north, a ptarmigan has to
moult three times between June and September; that is, from
winter to summer plumage, from summer to autumn, and from
autumn to winter white again. It is another interesting parallel
between tundra and dry desert that Zaroudnoi (37) noticed that
in Trans-Caspia the larks moulted so suddenly and so completely
that they were often unable to fly.

The moulting of the geese on the Yenisei tundras is accompanied by a partial migration. Large flocks appear at the beginning of June, travelling along the great river valleys. When
they reach the estuary the old birds at once go inland to breed
beside the little lakes which are dotted over the tundra; but
the birds of the previous year, which are not yet sexually
mature, remain in bachelor comfort on the river flats for another
month. Then for some days a constant procession of geese
passes up to the nesting grounds; and young and old undergo
the moult together, beside the pools where they themselves
were reared, among the month-old goslings of the season.
Ducks and geese normally shed their flight quills all at once,
not in pairs like other birds, and so for a time they are quite
helpless. The Samoyedes organise great battues at this time
and kill large numbers of geese. Trevor-Battye (33) has given

us an interesting account of one of these massacres on Kolguev; and I have described elsewhere a goose hunt on a smaller scale on the Yenisei (14).

It is not easy to explain this moult migration. The geese require a sanctuary during their helplessness, but there are plenty of places apparently as suitable as the already over-crowded breeding ponds. It is tempting to suggest a more subtle instinct. Hudson (15) refers to the curious habit of the semi-wild horses of the pampas, which, when dying, return to the human habitations they avoided when in health. He sup-poses that the animals' moribund condition reproduces in some measure the weariness and thirst of working days; and that the stable is associated with relief from bit and bridle, and with food and rest. It is conceivable that for the geese the critical time of the moult is ushered in by sensations of vague dis-comfort and restlessness which impel the birds to shift their ground; and if memory at all be theirs, the spot in which they were reared would of all others suggest warmth, shelter, and protection. Hudson compares the action of the dying horse with the statement of Darwin (10) that the huanacos of the Santa Cruz region in Patagonia, when about to die, seem to resort to certain bushy places close to the river. In one spot Darwin counted between ten and twenty skulls, and he observed that the animals must have crawled thither before death. He re-marks that he heard of similar places on another Patagonian river, and he noticed that wounded huanacos invariably walked towards the water. It is possible that the moulting geese and the dying huanacos in their extremity yield to the association of ideas which impels them towards those surroundings that formerly afforded them refreshment and rest. Perhaps it is an analogous instinct which causes a wounded horse or deer to crop herbage immediately after the accident, or a bird, anxious for her brood, to feign to pick up food or sip water.

The subject of bird migration is too wide to be dealt with here. Those who desire a discussion of the matter, with refer-ences to the literature, will find it in Coward's admirable little manual (8) or in Eagle Clarke's more exhaustive treatise (5). How migratory birds direct their flight, or what initiated the evolution of their long biennial journeys, we do not know, but it is safe to say that, as a general rule, a bird nests in the coolest part of its annual range.

Migration is not entirely a question of food. For example, in Britain some birds are migratory, while others of the same species remain resident throughout the year. Many birds leave their summer haunts while food is still plentiful, and in some cases the parents leave before the young ones, or *vice versa*.

The birds of the polar basin begin their southward migration very early. Already by the end of August our coasts are invaded by parties of waders, generally young birds of the year which are seldom accompanied by mature companions. It is a romance of geography to reflect that a bird, no bigger than a thrush, paddling over the flats of Norfolk or of Thanet, first saw the light scarcely two months before beside a glacier on Spitzbergen, or amid the lichens of the Taimyr; and that he regards our shores merely as a convenient caravanserai on the way to Sierra Leone or the Cape. Newton [25] gives two striking instances of great northern divers which were captured, one in Ireland and the other in the Faroe Islands. The Irish bird had a copper arrowhead embedded in its neck; the Faroese bird still carried a bone dart under the wing. These were reminders of a summer spent on the coasts of the Eskimo.

Seebohm suggested that the desire for light was at the root of the migratory instinct, and other writers have put forward variations of this idea. The most recent is that of English [12], who suggests that the longer polar day permits larger broods to be reared. He points out that it is generally admitted that birds in the tropics as a rule lay half the number of eggs of their allies in the north. In the tropics the hours of light and darkness are nearly equal throughout the year, and a tropical bird has only twelve hours in which to seek food for its young. Within the arctic circle at midsummer there is no darkness at all, and a bird, in northern Norway for example, can and does forage for eighteen hours at a stretch. Newly hatched chicks require a constant supply of food, and the parents work at high pressure[1]. It follows that a bird on the equator will only have half as long for food hunting as its relations in the arctic, and so will be able to provide adequately for only half the number of young. English further pointed out that certain nocturnal birds in the tropics, such as the oil-birds of the Trinidad and Venezuelan caves, which are related to the nightjars, lay clutches which are larger than those of the northern Caprimulgidæ, and he inferred that as the hours of darkness were longer, there was time to bring food for a larger family. However, other tropical nightjars, such as *Caprimulgus nigrescens* of Guiana, lay only one egg, whereas the common European goatsucker lays two, and there are also other facts difficult to reconcile with the theory. For instance, according to Feilden, Manniche, and others, the Greenland hare produces seven and sometimes eight young in a litter, while the British hare seldom has more than four. There is no question here of prey which can be captured only in daylight. The hare is herbivorous and suckles her young, and it is simply a matter of how much food the parent herself can assimilate, which, being semi-nocturnal, she can do irrespective of light or darkness[2].

However, even if this theory does not cover all the facts, there is no doubt that larger broods are the rule in the north, where, for example, the usual number of eggs for waders is four, for finches five or six, and for crakes and rails seven or eight; while in the tropics members of the same families usually produce clutches of two, two or three, and three eggs respectively. The arctic thrushes apparently also average a slightly larger brood than those in temperate regions. The clutch of the Greenland wheatear is said to be larger than that of the British bird; and even in northern Britain the meadow pipit may lay six or seven eggs, while in the south it is rare to find more than five (35).

The most notable exceptions to this rule are certain tropical woodpeckers which produce families as large as those of their

---

[1] Even in the midnight sunshine the birds take a kind of siesta. On the Yenisei I noticed that between midnight and two or three a.m. there was a general slowing down of activity, and many species took covert for a while.

[2] Beebe (*Zoologica*, vol. VI, no. 1) remarks that certain tropical mice produce small litters.

congeners in the north. English ingeniously suggests that tropical woodpeckers have an abundant and unfailing food-supply in the termites which are found so plentifully in the forests; and he supposes that, thanks to this, the birds are able to provide for a large brood in spite of the shorter hours of daylight. He compares the tropical woodpecker, tapping a termite gallery and collecting the insects that literally pour out, to a man who, desiring a bath, enters an up-to-date bathroom and obtains plenty of hot water by merely turning a tap. On the other hand, the northern woodpecker, which picks up insects here and there, is like a less fortunate bather, who must first draw his water from the well and then boil it up in a kettle over a fire of sticks.

Thus food and light alone cannot account for migration, and neither, so it seems, can temperature; for there are plenty of places between the equator and the arctic circle where a bird could find a cool climate without travelling so far north. Actually this may have happened in a few cases. Two chiff-chaffs, a genus which performs long annual journeys, are resident throughout the year in the Canary Islands, and are sufficiently differentiated from the forms on the continent of Europe to require sub-specific rank. The redstart-flycatcher, whose nearest relations on the mainland of North America are migratory, seems comparatively recently to have become domiciled in the island of Dominica in the West Indies. The vitelline warbler is restricted to the Grand Cayman and to another small island in the western Carribean. A similar species, the prairie warbler, which, although of rather smaller size, resembles the vitelline warbler in appearance and habits, is found in summer in the United States, and in winter migrates through the West Indies, including the islands where its congener is resident (21).

Sometimes certain individuals of a migratory species fail to go north in summer. The spotted sandpiper breeds in arctic America and winters in Guiana and Brazil; but in spring a few birds linger behind. They never breed, but loiter throughout July and August on the palm beaches and mangrove creeks of the Amazons and Orinoco. In the same way, little parties of turnstones and purple sandpipers remain on the shores of

Britain in summer. These stay-at-homes are generally birds of the previous year which have not yet attained sexual maturity.

As might be expected from the physical conditions, over half of the birds of the tundra are aquatic or littoral. The waders have pride of place in respect of numbers, voice, and colour in a region where songsters are scarce and sober plumage prevails. The nuptial dress of some of the plovers and sandpipers is as rich in hue as it is transient. Every tint, from yellow, chestnut-red, steel-grey, and warm buff, to contrasted black and white, is to be found among the grey and golden plovers, bar-tailed godwit, ruff, knot, sanderling, curlew-sandpiper, and phalaropes for two or three weeks between their arrival and the onset of the moult. Perhaps their splendour gains somewhat from its sombre background; but even in a setting of blue ice or sunlit snow, the phalaropes glow ruby-red, and the golden plover's chequered black and white *semé d'or* will bear comparison with the brilliant plumage of many tropical species.

The waders' love calls often provide the only bird music in the arctic wastes. Manniche thus describes the song of the knot on the stony uplands of north-east Greenland:

> The male suddenly gets up from the snow-clad ground, and producing the most beautiful flute-like notes, following an oblique line with rapid wing strokes, mounts to an enormous height, often so high that he cannot be followed with the naked eye. Up here in the clear frosty air, he flies around in large circles on quivering wings, and his melodious far-sounding notes are heard far and wide over the country, bringing joy to other birds of his own kin. The song sounds now more distant, now nearer, when three or four males are singing at the same time. Now and then the bird slides slowly downwards on stiff wings with the tail feathers spread; then again he makes himself invisible in the higher regions of the air, mounting on wings quivering even faster than before.
>
> Only now and then the observer—guided by the continuing song—succeeds for a moment in discerning the bird at a certain altitude of flight, when the strong sunlight falls upon his golden coloured breast or light wings.
>
> Gradually, as in increasing excitement he executes the convulsions of his wings, his song changes to single deeper notes—following quickly after one another—at last to die out while the bird at the same time drops to the earth on stiff wings strongly bent upward.

PLATE VII

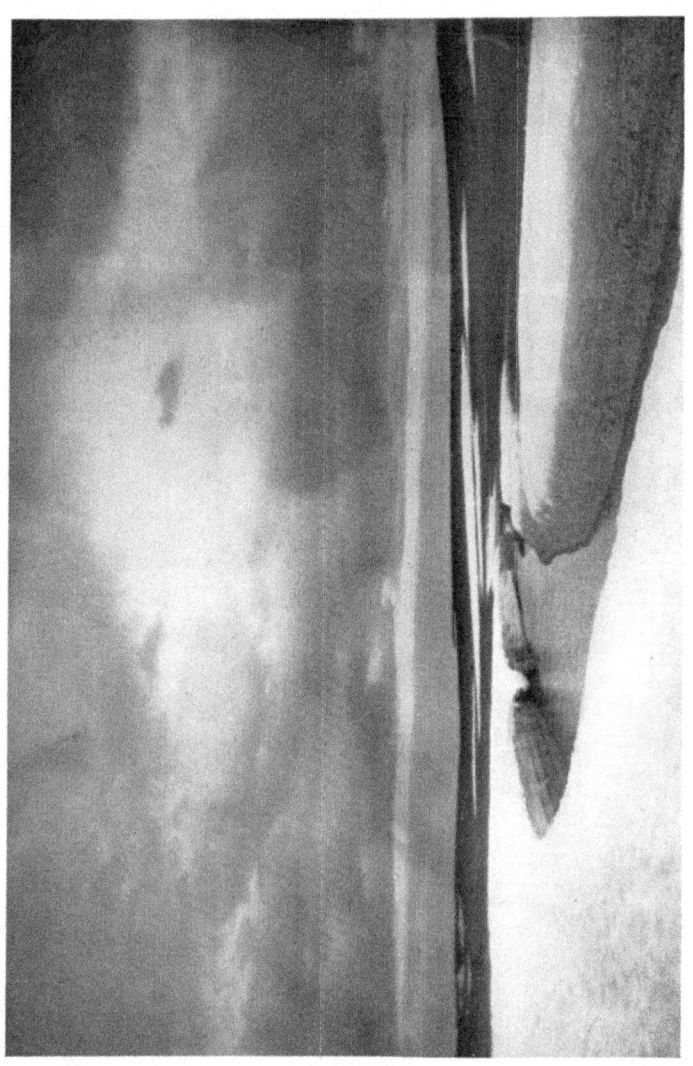

SUMMER ON THE TUNDRA

This fine pairing song may be heard everywhere for more than a month at the breeding places, and it wonderfully enlivens these usually desolate and silent regions.

On the marshy tundras of the Yenisei the melodious trilling of Temminck's stint, uttered as the bird hovers aloft, and the piping of the grey and golden plovers, like "horns of Elfland faintly blowing," are the most characteristic sounds. Besides these, in every morass and tarn, the lively voices of dunlin, stints, and phalaropes blend into a continual twittering murmur in such harmony with the character of the country that the spring songs of thrushes and warblers are never missed; and now and again the air thrums to the stately wing-music of a flock of wild swans.

The Passeres form only about twenty per cent. of the avifauna of the Yenisei tundra; but œcologically they are of interest because they are restricted in most cases to an environment which is the nearest possible to the typical environment of their kind elsewhere. Thus at the first delta, the Breokoffsky Islands, where the willows grow in a tangled mass a few feet in height, are found such representatives of a woodland avifauna as a willow-wren, the dusky thrush, the mealy redpoll, and an accentor. In Siberia the white wagtail is found far beyond the limits of forest growth; but, according to my observations, not in large numbers and always near running water, preferably close to human habitations, where the graceful little bird trips over the roofs of the huts, just as the pied wagtail visits the eaves of houses in this country.

The bluethroat does not nest on the open tundras; but in August small flocks of the birds of the year wander further north and visit the willow scrub along the river banks. This species belongs to the forest zone, and winters south of the Mediterranean. Lynes (22) records that in February, beside a crater lake in the Sudan, a little patch of rushes "scarcely equal to half of a lawn tennis court" held a bluethroat, a sedge-warbler, and two red-throated pipits. He remarks that it was perhaps "a mixed party from Lapland."

The red-throated pipit is a songster of the blood-royal of the skylark, and, like a lark, haunts sunny open places where there is some herbage. On the other hand the wheatear and the shore-lark come of stock which inhabits dry desert ground elsewhere, and on the tundra they are confined to the hill tops where the soil is bare and tolerably dry. What they find to eat in these deserts of deserts, unless perhaps Collembola, I never could ascertain.

The two passerines which do not conform to the environmental rule are the snow and Lapland buntings. In Britain, the snow-bunting is a bird of the mountain top, like the ptarmigan or dotterel; but as it ranges northwards it substitutes latitude for altitude, and nests freely in swamps and low ground, especially near heaps of driftwood. The Lapland bunting, which is the common perching bird of the Yenisei tundra, is catholic in its haunts, and nests anywhere, from a hollow log on the beach to a cleft in the earth of a mud-bluff.

The taste for the characteristic surroundings of their family does not belong to perching birds alone. The marshes are full of stints, ruffs, gulls, terns, and long-tailed and pintail ducks, just as the English bogs are occupied by redshanks, snipe, and mallards. The lichen tundra, which is the counterpart of high moor elsewhere, is the resort of golden plovers and curlew-sandpipers; and the dotterel and ptarmigan breed on the stony ridges of the river terraces, which are the Alps of the tundra.

In Scotland, the red-throated diver rears its young beside small moorland tarns, and on the Yenisei these birds breed in the little tundra lakes which satisfy their traditional requirements. This choice really jeopardises their brood, for they cannot lay their eggs until the ice has thawed late in July. Incubation is long, and consequently the young are scarcely fledged by the end of August when the weather breaks up. Divers are among the last birds to leave for the south, and in September 1914, when travelling through the Vaigach Strait, off the coast of Novaya Zemlya, I watched a constant stream of divers pouring southwards for half a day towards the continent of Europe—an unforgettable sight.

On the Yenisei the divers soon bring their young from the little breeding tarns to larger pieces of water where fish is plentiful. The adult diver is almost incapable of progression on land, but the young seem to possess this power in early days. I have elsewhere (14) described a diver chick which, though only a day or two old, used its legs and wings in a quadrupedal manner to cross swampy ground; and which, when placed in water invariably swam ashore and crawled into the grass. Trevor-Battye (33) records a like instance on Kolguev. The ability of the little diver to move on land is comparable to the diving habit of the young hoatzin (see p. 50) and probably enables it to follow its parents to fresh feeding grounds.

Birds of prey, peregrine falcons and rough-legged buzzards, also choose the ancestral site; and as cliffs and trees are not available, they build their eyries on the summits of low mudhills.

The rapacious skua, the *tchornaya tchaika* of the Siberians, is a bird of tundra and low moorland. These pirates will devour anything, from berries to young birds and small mammals, and they harry gulls and terns for the sake of the fish they have captured. I have seen a pair of arctic skuas attack a pintail duck which was in moult; and according to Manniche the long-tailed skua is a match even for the gyr-falcon. A third species, the pomatorhine skua, may be found far from land in the Polar Ocean. When temporarily ice-bound in the Kara Sea, about lat. 76° N., long. 62° E., winter having already set in, I saw a pomatorhine skua fly down from the north and linger for an hour or two round the ship. The skua's strength and ferocity make it a valuable ally on occasion. Popham (29) records that on the Yenisei he found the breeding grounds of some arctic skuas shared by a pair of bar-tailed godwits, who probably appreciated the protection against gulls and foxes. The bar-tailed godwit, one of the handsomest of waders, is among the first migrants to reach the tundra in spring. It breeds on the higher ground where the snow thaws early; and the Samoyedes say that the *tufek*, as they call it, runs round the frozen pools, tapping impatiently with its long bill and crying for the ice to melt.

Some of the wading birds of the tundra, such as the ruff, dunlin, and ringed plover, also breed south of the arctic circle, but others are natives of the high north, and are known in Europe only on migration. The handsome little curlew-sandpiper breeds only on the tundras of Siberia from the Yenisei eastwards, but its winter wanderings include half the globe, from the Cape to India, Malaya, and even to Australia and New Zealand. This disparity between winter and summer range is found in several arctic birds. Cooke has shown that the Asiatic golden plover occupies a breeding area of about seventeen hundred miles from east to west; but in winter it ranges laterally over a distance of ten thousand miles.

The curlew-sandpiper visits Britain in small numbers in spring and autumn; but some other species occur at one or the other season only, and evidently follow different routes on their northward and southward journeys. The white-billed diver apparently travels to its breeding grounds on the Taimyr along the valley of the Yenisei, but is unknown on the river in autumn. The little stint, the smallest wading bird, is commoner on British coasts as an autumn than as a spring visitor. It breeds on the tundras of the Old World from the White Sea eastwards; but its relative, Temminck's stint, is an unusual passage migrant to this country at any time, though its nearest breeding grounds are no further away than Scandinavia.

The most notable instance of a bird which follows a different route in spring and autumn, and traces annually a vast triangle, is the American golden plover, whose migration has been studied by Cooke (7). This plover breeds in North America from Alaska to Hudson Bay. In autumn the birds move leisurely into Labrador and Nova Scotia, whence, after resting awhile, they take wing and perform one of the most wonderful non-stop flights in the world—twenty-five hundred miles south to the South American mainland. Under adverse conditions some of the birds put in to the Bermudas or Antilles, but if the wind is favourable, they pass over without alighting. They rest and feed on the hot muddy beaches of Guiana, and then travel down the coasts of Brazil to Argentina and Patagonia, where they winter. The spring route lies more to the westward, across the equator to the Panama Isthmus, from Yucatan to Texas, and thence by the Mississippi Basin and the eastern line of the Rocky Mountains to the arctic breeding grounds. The annual journeys of the golden plover of the Old World are equally extensive but less well known. The European golden plover breeds in arctic and sub-arctic regions, from Iceland to the Yenisei, wintering in Africa. The Asiatic golden plover nests on the tundras of Siberia from the Yenisei to Alaska, and winters in India, Malaya, Australia, the Sandwich Islands, and Chile. In short these plovers are at home on the shores of every sea in the world. The same may be said of the grey phalarope, which ranges from the high north of both hemispheres to the south Pacific and Atlantic Oceans. This bird, and its ally the red-necked phalarope which breeds as far south as Ireland, are common on the Siberian tundra. They are distinguished from other waders by their palmately lobed feet, and the close duck-like texture of their plumage. They swim well and buoyantly, and in winter are sometimes encountered at long distances from land, riding on the waves with ease. The phalaropes in nuptial plumage are brightly coloured little birds and have attracted a good deal of attention because the relations of the sexes are peculiar. The female, which is rather the larger and more brightly coloured of the two, takes the initiative in courtship; and, having laid her eggs in a nest previously built by the male, in most cases she leaves incubation to her mate, who also undertakes the care of the young.

Recently, Morgan compared the coloration of the sexes in the phalaropes to that in the Sebright bantam. In this breed of fowls the cock is coloured like the hen, and it has been shown experimentally that this is probably due to an endocrine secretion of the testes, because after castration the cock develops plumage with the distinctive secondary sexual characteristics of other male fowls. The testis was found to contain cells resembling the luteal cells of the ovary, and the suggestion is that these cells produce a substance which inhibits the development of male plumage of the ordinary type. But lately Morgan [24] and Yocum [36] independently investigated the testis of the phalarope, and have failed to find luteal cells. Thus the controlling factor is still in doubt, though it may be remarked that the two cases are not quite parallel, since in the phalarope the change is reciprocal. Not only does the cock wear "hen" plumage, but the colour of the hen is that of a cock.

The male phalarope is a devoted parent, and when the broods are a few days old several families form a flock and the old birds unite in their defence. As evidence that food alone cannot account for migration, it may be mentioned that although on the tundra the grey and red-necked phalaropes inhabit the same marshes, and apparently depend on the same food, the former leaves the country nearly a month earlier than the latter.

The knot and the sanderling are birds of the stony rather than of the lichen tundra, and Manniche (23) has given a detailed account of their nesting habits. They visit Britain in considerable numbers in the spring and autumn, and breed in the uttermost fastnesses of the north, in the Taimyr, Greenland, and arctic North America. Their winter range extends to South Africa, Australia, and America down to Patagonia.

But an even greater wanderer than these is the arctic tern, which breeds from the latitude of Holland northwards to the circumpolar limit of land. In autumn the terns pass down the coasts of Africa and Australasia, and range into the Antarctic Ocean. In spring they return, arriving in June for a two months' sojourn among the swamps and snow-fields of the high north. It is one of the mysteries of ornithology that this great flock of birds, passing twice a year along ocean highways whose length outstrips that of two continents, is seldom if ever seen on passage between the arctic and antarctic seas. Cooke estimated that their annual journeys cover no less than twenty-two thousand miles; and that as they spend about fourteen weeks at their breeding grounds, and the same time or perhaps a little longer in their winter haunts, they must accomplish the round trip in about twenty weeks. This means an average flight of about one hundred and fifty miles a day, not reckoning the turns to and fro in chase of food. This estimate is perhaps rather liberal, but even so the migrations of the arctic tern are without parallel. As Cooke points out, this bird enjoys more light than any other creature. In the arctic its summer sun never sets, and perpetual daylight reigns in its winter haunts also. The greater part of its life is spent on the wing between sky and water; and with the solstices for its time-piece, its calendar is set to the revolution of the world.

It is a far cry from the arctic tern to the stay-at-home ptarmigan and willow-grouse. On the arctic mainland these birds move south irregularly in winter. On the Siberian tundras they linger until late October or November when darkness settles down, and then migrate to the forests. But on Spitzbergen and other islands the ptarmigan winter in their breeding quarters, and Trevor-Battye (33) records that willow-grouse furnish the chief winter food of foxes on Kolguev. These birds are so hardy that Manniche found in north-east Greenland that they returned as early as February, before the sun appeared; and, regardless of temperature, sought the open plains which had been swept bare by the wind.

Thus the vegetarian grouse are the only birds able, even partly, to face the rigours of the arctic winter, but it is the very intensity of the cold and depth of the snowfall that make this possible. The drifts afford shelter and warmth, and preserve plant life; and the same blizzards which pile up the snow sweep the country bare and uncover the buried food. It is not the herbivorous lemming and willow-grouse which feel the full terrors of famine, but the flesh-eating wolf and predatory owl.

Many people are acquainted with the exhibit in the entrance hall of the British Museum of Natural History, which illustrates the harmony of the coloration of arctic animals with their surroundings. The change to white is undeniably protective, and yet in the darkness of the northern winter the need for such protection must be less than is commonly supposed. The arctic fox is dimorphic, and both white and grey winter pelage is known. The rarer grey variety is the valuable "blue fox" of commerce; and it is interesting to notice that this grey winter coat approaches the colour of the young and of the summer pelage of the "white fox," which is greyish with a dark stripe—the so-called "crossed fox" of fur-traders. Both blue and white varieties seem equally successful in the struggle for existence, and in Iceland it is said that all the foxes are "blue." The hare and ermine, which live more above ground, have a seasonal change, though again, in the twilight the need for complete harmony cannot be very great. The male snowy owl, and to a ess degree the female, are nearly white all the year round. As this powerful bird has no enemies its colour cannot have evolved for defence; and it is doubtful whether it conceals its owner from the prey either, since in the half light it is not more obliterative than brown or grey, and in the summer a snowy owl against the dark tundra is as conspicuous as a sea-gull. In short, it seems as if the winter-white colour scheme originated in the sub-arctic region where the daylight is less curtailed, and has partly lost its function in the long northern night.

An interesting point in the arctic wading birds is the frequency of chestnut or pinkish underparts in their summer dress. A list of twelve species which nest in the Yenisei tundras contains 33 per cent. of chestnut-breasted forms, while only twelve per cent. are found in Britain. The difference would be even more striking if the list was extended to the whole circumpolar region to include the sanderling, knot, and the American red-breasted and buff-breasted sandpipers. Why the nuptial plumage should tend to be red is not known. The dress acquired immediately after the breeding season is usually grey or brown and white, which harmonises admirably with the foam of the sea shore or the sandbanks and mudflats of the winter quarters.

Some experiments made at the New York Zoological Park [2] suggest that the seasonal colour changes, and moults of birds are due to physiological states directly affected by environmental conditions such as food-supply and light. The scarlet tanager has red nuptial plumage, and moults into a green dress in the autumn. In the summer some of these tanagers, which were in full song but had not been allowed to breed, were put in cages with subdued light and fed on fattening food. As winter came on they ceased to sing, but they remained fat and in good health and they did not moult. During the winter it was found that if some of the birds were subjected suddenly to a much higher or a much lower temperature, they immediately "went off their feed" and began to moult into the green winter plumage. Other individuals were kept under these experimental conditions all winter and not brought into full light until the following spring. Birds treated in this way began to moult directly into the red summer plumage without passing through any intermediate green stage.

Similar experiments on a species of bobolink gave much the same result, and we are led to the conclusion that, in some kinds of birds at least, the character of the moult is determined by the environment. But so far the experiments have been made on very few species, and the explanations that suggest themselves will not altogether cover the state of things in nature. For instance, if darkness inhibits the moult, what are we to make of the arctic ptarmigan which moults three times a year, just when there is no darkness at all?

CHAPTER III

THE insect fauna of the tundra strikes the traveller in the first place with disappointment, in the second with wonder —with disappointment at the scarcity of individuals and species, with wonder that in surroundings so harsh and inclement such varied forms of high organisation can exist at all. The insects have at most ten or twelve weeks of activity. The remainder of the year must be passed in a resting stage of one kind or another, and it is the capacity to lie dormant in the cold season which determines whether or no a species shall range into the tundra region. Hence the Exopterygote orders, the Hemiptera, Orthoptera, and most "Neuroptera," start handicapped because there is but slight natural resting period in their post-embryonic development. It is a little surprising that so few of them are able to winter in the egg stage, for the steppe lands to the south, which in summer are a paradise of grasshoppers and bugs, experience a long and bitter winter with temperatures far below zero; but the limiting factor in the high north is not so much the cold of winter as the short cool summer and the wet soil. Grasshoppers as a rule oviposit in places that are not too damp, and the tundra earth is always chilled and saturated.

The disparity in numbers between the Exopterygota and Endopterygota is partly because the latter is actually the larger division. Thus Sharp (32) estimates the approximate numbers of the species for the whole world as follows:

Hymenoptera (bees, wasps, etc.)	...	30,000 species or	10 %
Coleoptera (beetles)	...	150,000 ,,	50 %
Lepidoptera (butterflies and moths)	...	50,000 ,,	16 %
Diptera (flies)	...	40,000 ,,	13 %
Hemiptera (bugs)	...	18,000 ,,	6 %
Orthoptera (grasshoppers, etc.)	...	10,000 ,,	3 %

These figures may serve as a standard for comparison, though the proportions are probably misleading, for the numbers of Hymenoptera and Diptera, groups which have hitherto attracted less attention than the rest, must be higher than those given. Indeed there is some evidence to show that in the arctic the flies, which have the most complete metamorphosis of all Endopterygota, are the dominant family.

The first comprehensive catalogue of the insects of an arctic area is that of Zetterstedt(38) who gave the families in order of numbers of species in Lapland as follows:

Diptera	...	1245 species or	35 %
Coleoptera	...	1001 ,,	28 %
Lepidoptera	...	499 ,,	14 %
Hymenoptera	...	426 ,,	12 %
Hemiptera	...	232 ,,	6 %
Neuroptera or "net-winged" insects	...	123 ,,	3 %
Orthoptera	...	14 ,,	0·3 %

It is interesting to compare the figures for Hymenoptera, Lepidoptera and Hemiptera, which have much the same proportion to the total whether we take the World Fauna, according to Sharp, or the Lapp Fauna, according to Zetterstedt.

In the *Fauna Arctica* of Romer and Schaudinn, the orders are represented as follows:

Coleoptera	714 species
Diptera	358 ,,
Hymenoptera (excluding sawflies)	565 ,,
Sawflies	228 ,,
Lepidoptera	797 ,,

But these figures have not much comparative value because the butterflies include many species from the forest zone, while the list of beetles is compiled more exclusively from the tundra region. De Meijere (11), who catalogued the flies, restricted himself to records from the arctic islands (including Greenland). He remarks that flies do not form a large part of the arctic insect fauna, for only twenty species are known from the islands west of Greenland, and only nine species from the New Siberian Islands. But even so, the list is by no means insignificant, and judging from other orders, it would probably be doubled if mainland forms were included. Two-thirds of the Hymenoptera catalogued are Scandinavian, and of these only eighty-six are strictly tundra forms. The list of sawflies is relatively long, but again only twenty-nine species have been recorded from the arctic islands.

The most satisfactory figures for comparison are those given by Jakobson (16) who, from his own collections in Novaya Zemlya, and from the records of other observers, arranges the insect orders of these sterile and unforested regions in orders of numbers as follows: Diptera, Hymenoptera, Coleoptera, Lepidoptera, Collembola, "Neuroptera," Rhynchota (Hemiptera), Orthoptera. He remarks that flies seem to be the dominant group in the north, and points out that only in Iceland are they outnumbered in species by beetles and bees[1].

That there is a regular sifting out of the different forms is readily seen by comparing the total insect fauna of Norway and Sweden with that of the strictly arctic part of Scandinavia. As we proceed northwards, first the grasshoppers, and secondly most of the bugs disappear. The "Neuroptera" (caddises, mayflies, dragonflies), especially those with aquatic larvæ, fight the climate for a little longer, and then the battle with the environment is left to the four higher orders, with the addition of the Collembola, which are themselves the most specialised of the Apterygota. Within the orders, certain families are better adapted than others to face conditions in the high north. Among flies, for example, the Tipulidæ (craneflies) and Chiro-

---

[1] Later collections from some of these regions have brought to light new forms. For instance, recent Danish writers include 175 species of Diptera on the Greenland list; but Jakobson's figures, as given here, hold good for comparative purposes. The insect fauna of the arctic lands is certainly larger than our present knowledge of it.

nomidæ (midges) are the dominant families of Nematocera, and the Anthomyzidæ, Heteromyzidæ, and Syrrphidæ among the Brachycera (16). The tundra beetles belong chiefly to the carrion and carnivorous feeding families, Carabeidæ, Dytiscidæ, and Staphylinidæ: vegetable feeders, such as the weevils, are poorly represented.

Beyond the forest limits there are no wasps, and no bees with the exception of humble-bees; and save for a few species of Ichneumonoidea, the great division Hymenoptera-Parasitica is unrepresented on arctic islands.

Distribution of the orders of insects in the arctic.
(After Jakobson)

	Novaya Zemlya	Spitz-bergen	Islands west of Green-land	Green-land	New Siberian Islands	Iceland	Total
Coleoptera	22	1	3	35	8	92	161
Hymenoptera	48	18	13	54	2	75	210
Diptera	100	54	25	94	9	78	360
Lepidoptera	12	2	23	43	—	51	131
Trichoptera	3	1	1	6	—	10	21
Neuroptera	—	—	—	1	—	1	2
Odonata	—	—	—	?1	—	—	?1
Plecoptera	3	—	—	—	—	2	5
Ephemeridæ	—	—	—	1	—	—	1
Orthoptera	—	—	—	—	—	—	—
Rhynchota	1	1	3	8	—	12	25
Collembola	16	16	5	15	—	5	57

The outstanding feature of the tundra Lepidoptera is the predominance of day-flying forms. This is what might be expected in lands which enjoy perpetual daylight in summer. Iceland, however, is an exception to this rule, and possesses scarcely any butterflies and few moths. The most northerly ranging forms belong to such genera as *Colias*, *Argynnis*, *Chrysophanes* and *Lycæna* among butterflies, and *Anarta*, *Plusia*, *Cidaria Hyphoraia* and *Penthina* among moths. More southern species are of the genera *Pieris*, *Chionbis*, *Hesperia*, *Arctia*, *Agrotis*, *Tortrix*, *Tinea*, etc. Sphingidæ, Bombycidæ and the genera *Zygæna* and *Sesia* are scarce on the tundra, but increase in numbers within the forest zone. Most of the tundra insects require two summers or more to complete their development, and tend on the whole to be smaller than their congeners to the south. They sometimes exhibit melanic variations, and there is frequently a thickening in the clothing of scales or hairs, which is specially noticeable in humble-bees.

Johansen (17) has observed that hibernating insects can survive temperatures of at least − 50° F., and when we remember that butterflies and humble-bees have been taken as far north as lat. 82° N. in Grinnell Land, it is clear that some forms must be able to resist cold greater even than this. This observer has supplied valuable records of how the different forms pass the winter in Alaska. Lepidoptera hibernate as larvæ, and perhaps as pupæ; and Coleoptera hibernate as larvæ, pupæ, or imagines. Some Hymenoptera, such as sawflies, winter as larvæ or pupæ, but the queen humble-bee hibernates in the imaginal state. A few Diptera also hibernate as adults, but those with aquatic larvæ and pupæ pass the winter in submerged mud, and the same may be said of the Neuroptera.

The relations of tundra flowers and insects would repay further study. Pagenstecher (26) quotes Vanhoffen to the effect that in Greenland bees and butterflies seldom visit flowers, but this is not altogether in agreement with the observations of other travellers. Aurivillius (1) has discussed the question at some length, and the result of his researches goes to show that the rôle of insects in pollination is progressively less as we advance north. Round the arctic coasts, green, yellow and white flowers, which are usually regarded as anemophilous, or at any rate less highly specialised for insect pollination, predominate over the red and blue blossoms which have been shown by experiment to be colours selected by bees. Moreover, the part played by flies is relatively greater, while that of bees and butterflies is smaller.

Aurivillius has called attention to *Pedicularis lanata* in Spitzbergen. This genus is specially adapted for fertilisation by bees, and Feilden (13) remarked that it was visited by humble-bees in Grinnell Land. Two species of *Pedicularis* are widely distributed in Spitzbergen and have been shown to set seed, but no humble-bees are known in the archipelago. Unless the plant is self-fertilised, in which case the elaborate devices to ensure cross-pollination seem superfluous, or, as is more likely, some other insect acts as pollinator, it is difficult to understand how *Pedicularis* maintains itself on Spitzbergen.

Percentage distribution of methods of pollination in the arctic.
(After Aurivillius)

	Schonen	Fin-mark	Iceland	Green-land	Novaya Zemlya	Spitz-bergen
Total number of flowers	1089	501	349	353	185	116
Anemophilous flowers	25	33	38	38	32	37
Entomophilous flowers	74	67	62	61	67	63
"Fly" flowers with open nectaries	36	41	52	51	59	53
"Bee" flowers	21	17	17	17	11	?8
"Butterfly" flowers ...	5	6	2	4	2	1
Other groups	37	39	27	29	28	17

Percentage distribution of insect- and wind-pollinated flowers in the arctic. (After Aurivillius)

	Schonen	Fin-mark	Iceland	Green-land	Novaya Zemlya	Spitz-bergen
Corolla or calyx red or blue	24	20	19	15	14	10
Corolla or calyx green, white or yellow ...	75	79	80	84	86	89

The scarcity of "butterfly" flowers in Iceland is clearly correlated with the rarity of day-flying Lepidoptera in that island; and Aurivillius points out that the six plants mentioned in his list belong to the genera *Silene*, *Lychnis*, *Orchis* (*maculatum*), *Plantanthera*, and *Viscaria*, which are known elsewhere to be visited by night-flying moths.

As a field for the œcologist the tundra is often overlooked in favour of regions with a richer flora and fauna, but actually it

is the paucity of species which makes it specially valuable to
the student of bionomics. The problems are simplified because
the action of one species on another is less obscured by the
influence of other forms whose bionomical standing is doubtful.
Nevertheless the web of life, even in a region so barren and
sparsely populated as the Siberian tundra, is complex enough,
as is shown by the accompanying diagram, which deals only
with a few of the higher forms, composing four principal bionomical complexes, each centred round a food-stuff. These
food foci are green plants, reindeer moss, mosquitoes, and lemmings.

Green plants, chiefly grass and willow, are the staple food
of lemmings, grouse, and geese. To a less extent other animals
rely upon them also. Reindeer browse on them, and they
provide covert for nesting birds.

The lemming itself is the focus of another complex as the
chief food of foxes, wolves, and birds of prey. If green food is
scarce and lemmings disappear, the immediate effect is not felt
by the carnivores, but by the birds whose eggs and young offer
alternative prey. Under such conditions, birds like the grouse
suffer twice over—first through lack of green food, and secondly
through the destruction of their broods.

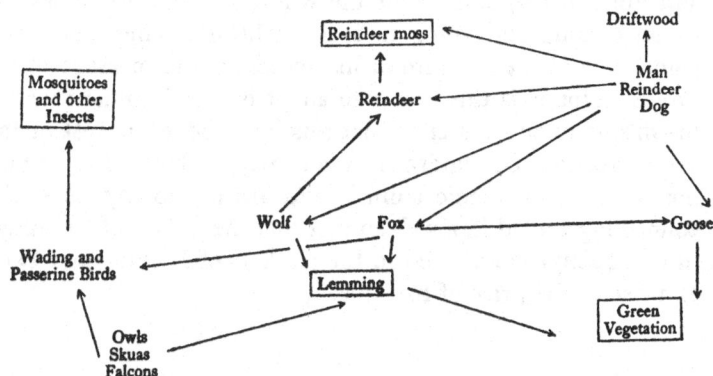

Mosquitoes and other insects are the focus of a great bird
complex; and here failure of the food-supply will affect various
forms differently. The perching birds will feel the pinch more

than the waders, which can feed alternatively on small aquatic animals.

The reindeer moss determines the range and abundance of the reindeer, and hence the movements of the wolf and of man.

The introduction of man into the scheme lifts it into the debatable land between bionomics and economics; and yet there is no option, for the Siberian hunter and herdsman is as dependent on the lichen pasture for his flocks as are the tundra deer themselves. In other respects he has risen higher and demands more of life. This "more" he pays for in foxskins, and as the numbers of the foxes fluctuate with the lemming population, so his standard of living rises and falls. The amenities of life become its necessities, and the prizes that he snatched in his progress grow into chains hampering his activities. The site of permanent settlements along the Yenisei estuary are determined by whether or no driftwood from the forest accumulates on the river beaches; for driftwood is the only fuel on the tundra, and without fuel man must migrate or perish in winter. Moreover his summer activities depend on the mosquito complex, and on fish and geese, for the torment of the first drives him and his reindeer to the highlands, while the other two draw him back to the rivers and marshes to fish and hunt in preparation for the winter. So his life is a continual coming and going to evade cold and hunger and discomfort; and the measure of his success is the measure of his progress, for it is this power to adapt his environment within his means to suit his ends that sets him above the rest of the brute creation. So tundra man is slowly evolving from a bionomic to an economic animal; and already to-day he is the connecting thread in a web so vast that the failure of lemmings in the Taimyr is reflected in the markets of London and Paris as a rise in the price of foxskins.

BIBLIOGRAPHY

(1) AURIVILLIUS, C. (1885). "Das Insektleben in arktischen Ländern." Nordenskiold's *Studien und Forschungen*, Bd. VII. Leipzig.

(2) BEEBE, C. W. (1908). "Preliminary Report on an investigation of the seasonal changes of color in birds." *Amer. Nat.* vol. XLII.

(3) BENT, A. C. (1921). "North American Gulls and Terns." *Bull.* 113, *U.S. Nat. Mus.*

(4) BORLEY, J. O. (1922). "The Plaice Fishery and the War." *Fishery Investigation*, series II, vol. V. London.

(5) CLARKE, W. EAGLE (1912). "Studies in Bird Migration." 2 vols. London.

(6) COLLETT, R. (1895). "*Myodes lemmus*: its habits and migrations in Norway." *Christiania Videnskabs Selskabs Forhandlinger*, No. 3.

(7) COOKE, W. W. (1911). "Our greatest travellers." *Nat. Geog. Mag.*

(8) COWARD, T. A. (1912). "The Migration of Birds." Cambridge.

(9) CZAPLICKA, M. A. (1916). "My Siberian Year." London.

(10) DARWIN, CHARLES (1891). "The Voyage of the *Beagle*." London.

(11) DE MEIJERE, J. C. H. (1910). "Die Dipteren," in Romer and Schaudinn's *Fauna Arctica*, Bd. V. Jena.

(12) ENGLISH, T. M. SAVAGE (1923). "On the greater length of day in high latitudes as a reason for spring migration." *Ibis*.

(13) FEILDEN, H. W. (1877). "On the Mammalia of North Greenland and Grinnell Land." *Zoologist*.

(14) HAVILAND, MAUD D. (1915). "A Summer on the Yenisei." London.

(15) HUDSON, W. H. (1903). "A Naturalist in La Plata." London.

(16) JAKOBSON, G. (1899). "Insecta Novaia Zemliensis." *Zool. Centralb.* (abstract in German), Bd. VI.

(17) JOHANSEN, FRITS (1921). *Report of the Canadian Alaskan Expedition*, 1913–18. Ottawa.

(18) JOURDAIN, F. C. R. (1922). "Birds of Spitzbergen and Bear Island." *Ibis*.

(19) KENDREW, W. G. (1922). "The Climates of the Continents." Oxford.

(20) KROPOTKIN, P. (1910). "Mutual Aid: a Factor in Evolution." London.

(21) LOWE, P. R. (1911). "A Naturalist on Desert Islands." London.

(22) LYNES, H. (1924). "On the Birds of North and Central Dafur." *Ibis*.

(23) MANNICHE, A. L. V. (1910). "The Terrestrial Mammals and Birds of North-East Greenland." Copenhagen.

(24) MORGAN, T. H. (1923). "On the absence of lutear cells in the testes of the Phalarope." *Amer. Nat.* vol. LVII.

(25) NEWTON, ALFRED (1896). "A Dictionary of Birds." London.

(26) PAGENSTECHER, ARNOLD (1902). "Die Arktische Lepidopteren Fauna," in Romer and Schaudinn's *Fauna Arctica*, Bd. II. Jena.

(27) PATTEN, C. J. (1906). "The Aquatic Birds of Great Britain and Ireland." London.

(28) PATTERSON, A. H. (no date). "Wild Life on a Norfolk Estuary." London.

(29) POPHAM, H. L. (1897). "Notes on the Birds observed on the Yenisei River in 1895." *Ibis*, vol. III.

(30) RAE, J. (1852). *Journ. Roy. Geol. Soc.* vol. XXII.

(31) SEEBOHM, H. (1901). "The Birds of Siberia." London.

(32) SHARP, D. (1918). "Cambridge Natural History," Pts. I and II, "Insects." London.

(33) TREVOR-BATTYE, AUBYN (1895). "Icebound on Kolguev." London.

(34) VON TRAUTVETTER, E. R. (1848). Middendorf's "Reise und Briefe," Bd. I.

(35) WITHERBY, H. F. (1920). "A Practical Handbook of British Birds," vol. I. London.

(36) YOCUM, HARRY B. (1924). "Luteal Cells in the Gonad of the Phalarope." *Biol. Bull.* vol. XLVI, No. 3.

(37) ZAROUDNOI, N. (1885). "Oiseaux de la Contrée Trans-Caspienne." *Bull. Soc. Imp. Nat. Moscou*, t. LXI.

(38) ZETTERSTEDT, K. (1840). "Insecta Lapponica."

PART IV

THE TAIGA

A BOOK of this character would be incomplete without some description of the climatic formation of coniferous forest which covers sub-arctic Eurasia and America. This wooded zone, which merges into tundras and "barren grounds" in the north, and into steppes and meadow land in the south, is so uniform, and at the same time so impressive in its simplicity, that it is surprising that it has received no definitive name in the many countries that it covers, except in Siberia. There the Russian settlers call the flat marshy forest that they live in, the "taiga[1]."

The Siberian taiga from the Pacific Ocean to the Ural Mountains is some 3600 miles long by 800 miles wide—the largest forest in the world—and yet its literature is insignificant. Its nearctic counterpart has claimed much more attention, and has even found a place in poetry and fiction. No Hiawatha nor Chingachgook has hunted in Siberia. The jungles of the Congo and Amazons are less known, but incomparably more has been written about them. The great forest which lies to the north of the Trans-Siberian Railway, along the valleys of the Ob, Yenisei, and Lena Rivers, has been explored. Fur-traders, lumbermen, hunters, fishermen, and prospectors for gold and graphite, have penetrated until there is little left blank upon the map; but these immigrants have almost always used the rivers as corridors for passage, and even then have seldom left records of their wanderings. It is not surprising that this should be so, for the taiga has relatively little to offer the scientific traveller or sportsman that they cannot obtain in greater measure elsewhere. Transport is difficult, the climate is extreme, the scenery is monotonous, and the flora and fauna are meagre in comparison with those of forests and savannas in other parts

[1] Morgan (9) states that this term is also used by trappers in the Altai for forested mountains where squirrels and other fur-bearing animals are hunted.

of the world. Hence the taiga has been overlooked, and its exploration has been at the instance of trade rather than of science. The present account, based on a short journey through the Yenisei taiga in 1914, can give no more than a sketch of the country and its environmental conditions.

In contrast to another type of woodland—tropical jungle— where a large number of different species of trees are crowded together, the taiga is a great assemblage of trees belonging to

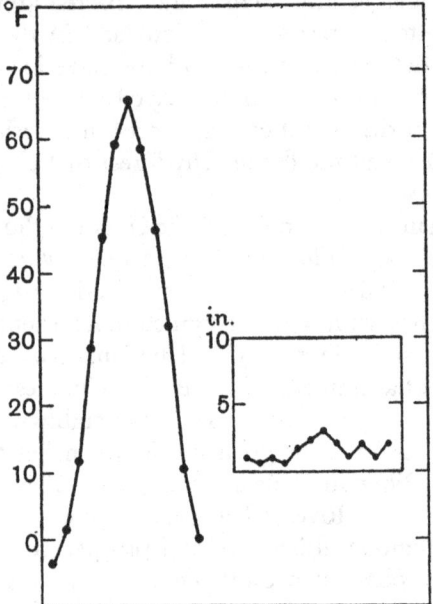

Fig. 8. Average monthly temperature and rainfall of Tomsk in the taiga region.

comparatively few species. These are mostly xerophilous in habit, since they grow in soil which is physiologically dry, and in winter they are exposed to bitter desiccating winds. The frost and snow begin in October, and continue until late April or May; so that there are only four or five months of the year when water is freely accessible to the roots. However, cold alone does not inhibit, though it retards, the growth of trees. Verkhoyansk has a mean January temperature which is actually below the winter mean calculated for the North Pole, and yet it is within the forest zone. Its summer mean is however comparatively high, for the maximum annual temperature range is 120°, as against about 75° at the Pole, and this is sufficient to permit the growth of trees. But even in summer in the taiga, absorption of water is retarded, partly by the coldness of the soil, and partly by the acidity of the humus which accumulates in the deep shade. As a matter of fact, parts of the sub-arctic zone are not ideal for forests at all. Protoplasm is ductile under circumstance, and does its best to maintain existence wherever it finds footing, but this does not mean that it would not flourish better elsewhere. That the climate is not really favourable to forest may be judged by Middendorf's observation that even in the southern part of the taiga the trees are never more than one and a quarter feet in diameter, and seldom more than a hundred and fifty years old[1]. The sitka spruce of North America grows much larger and more rapidly when transplanted to the mild climate of Scotland. Thompson Seton (11) calculated that timber, which reaches a certain size in ten years in Connecticut, would take three hundred years to attain the same girth within the arctic circle.

The taiga is a forest of mixed growth. In the north, larch, spruce, and birch predominate; but in more southerly latitudes they are interspersed with fir, pine, and the Siberian cedar, a valuable timber tree which sometimes attains considerable size. Here and there deciduous thickets of aspen, alder, willow and mountain-ash mingle with the conifers. A dry mountain pine wood, such as is seen in Scotland, gives no idea of the Siberian forest south of the arctic circle. If the taiga could be denuded of its trees it would appear as a low rolling plain, traversed by gentle ridges and shallow valleys. The hollows are often marshy, and contain a sluggish stream. These boggy places, which in the north are perpetually frozen beneath the surface, are covered with mats of hoary sphagnum and a few scattered spruce trees. They correspond to the "muskegs" or moss-bogs of the Canadian forests, and life there in summer is almost intolerable from the mosquitoes.

[1] This is no doubt partly due to the action of forest fires.

The lianas and climbers, characteristic of tropical jungle, are replaced by the thin wiry saplings of birches, alders, and rowans, which spindle up to heights out of all proportion to their girth in the race for sun and air. They grow as straight and close-set as the bars of a cage; and although often no thicker than the little finger, they make human progress difficult and slow. This deciduous growth usually occurs wherever the conifers have been destroyed at one time or another, either by man or by fire, and as the latter is frequent, these tangled patches are common.

Seen from one of the great rivers, the taiga is like the sea in its immensity and monotony. The individual trees are not very large, nor very thickly set, but they seem illimitable. The eye passes from horizon to horizon until the dull green of the nearer foliage fades into the blue distance. And if we could take wings the same view would be ours for hundreds, even thousands, of miles, for this great forest encircles the dry land of the world. The vastness of the taiga is realised most intensely at dawn. While the ground below is still in darkness, birds perch on the tops of the trees to await the approach of light. Suddenly an ousel, far away to the eastward, begins to pipe faintly; and as the minutes pass, another and another, nearer and nearer, join in the chorus, until every tree bursts into a pæan of song, which in turn is taken up by the expectant multitude in the forests to the westward. It seems as if the great wooded shoulder of the earth, rolling eastwards into the sunrise, awakes one songster after another, until Asia and Europe, from Pacific to Atlantic, are linked together by a chain of thrushes' music.

When the river bank is left behind, the traveller in the taiga plunges at once into a world of bewildering monotony and intricacy. The trees themselves are not very impressive, but their dark spreading branches shut out the light, and their sameness is such that, without close attention to the compass, there is every risk of losing oneself within a hundred yards of camp. The evergreen foliage carries no stamp of season, and the twigs and boles are hoary with lichen. Where a little sunlight can filter down, the ground is clothed with clumps of juniper, bilberry, raspberry, crowberry, and marestails; and in summer these are accompanied by a few flowers, such as *Pyrola*, *Hepatica*, strawberry, and cloudberry. Otherwise only mosses, lichens, and sometimes fungi, can grow. In the deepest shade even mosses are starved out, and the earth is smothered with a great muffling blanket of pine-needles, the leaf-fall of years. Where a tree falls, there it lies; and as you walk, you are apt to slip down, through sodden débris, knee-deep into the heart of a stump which, when standing, may have seen Yermak enter Siberia. Indeed the taiga is a veritable Golgotha of trees. In warmer climates bacteria and other organisms work quickly to dispose of dead timber, but in the gloom and cold of the arctic circle the dissolution of a tree is a lengthy process. Even the life of the standing forest seems slow and lethargic in comparison with the exuberant growth of tropical jungle. The shade and the silence are both oppressive. The sudden call of a bird, such as the goblin cry of the cuckoo, *Cuculus optatus*, is startling, and the crackle and rustle of one's own passage seems a desecration in that quiet place. After crossing a mile or two of this forest, stooping under branches, squeezing between saplings, and stumbling over broken boughs and mouldering stumps, it is a relief to emerge even into one of the moss-bogs, where at least one can see the sun and feel the wind.

Along the banks of the rivers the character of the forest is frequently changed by the tremendous spring floods, which carry away the conifers and make a space which is filled later by quick-growing deciduous trees, such as alders, hazels, and willows. These thickets are swept and inundated every year at the break-up of the ice; and early in June, before the new foliage has covered the ravages of the subsiding water, they present a most desolate appearance in the wet and windy weather that prevails at that season. The thin trunks of the trees are washed bone-white; and as the wind whistles through them, their naked limbs creak and clatter drearily. They are distorted by the strain of years of storm and flood, and in places their boles lie almost level with the ground, while the branches, combed and battered by floating ice, form an almost impenetrable barrier to human advance. For a height of ten or twelve feet they are festooned with a fantastic garnishing of flotsam—dry grass, twigs, and rushes—brought down by the floods. This mat-like covering is so thick that the explorer must force his way through it, as through a screen. All animal life is hidden, except when a colony of fieldfares is startled into harsh screams or a sandpiper flies away crying mournfully. On a stormy evening such a thicket, with its rattling branches hung with rushes that wave like dishevelled hair above the sucking, sighing water, suggests a *macabre* assemblage, not of trees, but of tortured sentient creatures. These willow forests sometimes extend for miles along the Middle Yenisei, over a series of islands intersected by channels which are themselves choked with fallen timber and impassable to a boat. A week or two later their aspect is more cheerful. A veil of foliage covers the naked branches; the waters recede from the sandy beaches, and hardy flowers, rhubarb, cresses, heartsease, and garlic, appear. These attract bees and other light-loving insects, which in turn bring insectivorous birds, such as wagtails, fly-catchers, and warblers, whose haunts are along the open river banks rather than in the dark conifer forest behind.

Brehm's (3) graphic description of the silence and apparent emptiness of the taiga is one that later travellers can only confirm. A few hundred yards behind the settlements the sound of man's occupation ceases, muffled by the trees. Human voices, the ringing of axes, and cackling of poultry, give place to stillness broken only by the chirp of a brambling or the tap of a woodpecker. The impression is that, compared with forests in more genial climates, animal life is scanty. Insect forms are neither striking nor very numerous, and reptiles and amphibians are few. Even in the forests of North America, which have been better surveyed than those of Asia, Preble (8) found only one species of toad, three of frogs, and one of snakes, in the Athabasca-Mackenzie region. Except for a few rodents, such as voles and squirrels, the mammals are hidden in the depths of the forest in summer; and although the list is long, including the bear, wolf, glutton, otter, ermine, sable, lynx, elk, etc., the advance of civilisation and the inroads of fur-hunters have reduced their numbers and restricted their range. This is the more to be deplored, because the mammalian fauna of the region

PLATE VIII

THE TAIGA NEAR THE ARCTIC CIRCLE

is a fine one, and includes some forms of large size. On the whole the taiga mammals are larger than the same forms on the tundra and steppe to north and south of them. Thus forest reindeer are larger than tundra deer, and forest wolves than steppe wolves. The lemming is, however, an exception to this rule[1].

Even bird life, which is comparatively plentiful, is readily overlooked. Drawn by the need for light, many species nest and feed at a considerable height from the ground; and, unless he shoots them, the naturalist knows them only as silhouettes against the sky. Even where the forest has been felled round the settlements, the denseness of the thickets of secondary growth makes observation difficult; and such places also lose much of their bird population after the thaw, when many species retire into the deeper taiga to breed. It is a popular misconception that the birds of the northern forests are dull of hue. Some of them rival tropical forms in brilliancy, and they have the additional advantage of a sombre setting. Conspicuous colouring in nature depends much on the background. A tropical parrot, which gleams in the hand like a jewel, may be invisible in its native trees, while a scarlet grosbeak in the dark taiga is as arresting as a flash of red light. The red-spotted bluethroat, pine grosbeak, and crossbill are other cases in point. The black, three-toed, and greater and lesser spotted woodpeckers are also conspicuously coloured black, white, and red; and the brambling, little bunting, siskin, waxwing, nuthatch, nutcracker and Siberian jay, though not so brilliant, are nevertheless pleasing in hue. Most exquisite of all are the mealy redpolls, which are pearl-grey and pink. In summer they haunt the tops of larch and birch trees, and their frosted-roseleaf plumage, against a lattice of blue sky and dark twigs, is one of the most beautiful sights in the forest.

The taiga also possesses some fine songsters. The bluethroat and Siberian rubythroat rival the nightingale, and the thrushes, of which there are a number of forms, including our own winter visitors, the redwing and fieldfare, are often

[1] For a discussion of the ranges of various mammals of the North American forest zone, see Allen (1).

musicians of the first rank. Typically, woodland birds, the world over, have loud voices, a character which they share with nocturnal forms such as owls and goatsuckers, for the reason that where dense covert or darkness hinders sight, a bird may find its way back to its fellows by sound. Conversely, birds which dwell in open country are often rather silent, or at least have a soft call. This is true of the Russian steppes, and according to Hudson (6) of the pampas also; while Beebe (2) remarks that even in the South American forest those species which frequent clearings have less powerful voices than those which inhabit deep jungle.

The swampy parts of the taiga support several species of wading birds. The common and Terek sandpipers and the oyster-catcher are found along the sandy beaches of the rivers; and the willow thickets and deeper woodland hold the green and wood sandpipers, both of which visit Britain on passage. In Europe the wood sandpiper often breeds on the ground; but in the taiga, like its congeners, the green sandpiper and the solitary sandpiper of America, it makes its nest in spruce trees, usually in old squirrels' dreys and fieldfares' nests. Three ducks of the forest zone, the smew, goosander, and goldeneye, likewise nest in trees at some height from the ground. The woodcock, dusky redshank, great, common, and pintailed snipe are all characteristic of these swamps. The latter has a curious love-song in the spring-time, comparable to the "bleating" of the common snipe. Rising to a great height, the bird dives vertically earthwards, and the air, rushing through the modified and stiffened tail feathers, produces a loud hollow roar which is audible more than a mile away.

Diurnal birds of prey, eagles, falcons, and buzzards, are common, and the taiga has a long list of owls, which are eminently suited to life in the dark forest. The common game birds are the capercailzie, black grouse, and hazel hen, which belong to a group whose headquarters are in North America, and the taiga along the Pacific slope is even richer in these forms. In winter the scales of the feet of the wood-grouse grow until the toes are surrounded by a horny fringe which acts like a snowshoe and supports the bird on powdery drifts and frozen branches. In the same way certain desert-living lizards in various parts of the world have developed fringes to the toes to prevent them from sinking in the loose sand.

Summer conditions in the taiga present no special problems. The temperature is not extreme, and food and water are comparatively abundant. The stress of life comes in winter when the thermometer sinks far below zero, and famine and drought, produced by the intense cold, may bring disaster.

Most of the birds of the taiga are migratory, but they do not winter so far south as those of more northern latitudes. In fact many of them may be termed vagrants rather than migrants, for they flock together and wander erratically about the borders of the forest, wherever food is still to be found. The central plateau of Asia, with its mountains and deserts, presents a barrier to any but strong-flighted species. It directs the current of migration from Western Siberia towards the Mediterranean, North Africa, and Mesopotamia, while the eastern stream moves into China, and along the Pacific coast[1]. On the whole, as might be expected, the warblers, thrushes, and waders, with other insectivorous, or semi-insectivorous birds, perform the longest journeys, wintering in Persia, Western India, China and the Malay Archipelago. The seed-eating finches, and also the tits and woodpeckers, which feed on bark-living insects, are vagrant in the southern part of the taiga itself, unless in very severe weather when they sometimes invade Western Europe. The best known of these sporadic irruptions are those of the crossbill and waxwing.

The crossbills are completely adapted to life in the conifer forest. They live in the upper branches of the trees like little parrots, which they resemble somewhat in their gay plumage, and the beak is specially modified to split open fircones. The upper and lower mandibles are crossed so that the tips overlap, and the

---

[1] Prejevalsky (9) remarks that the autumn migration in the Tian-Shan Mountains is not very striking. However, he gives a vivid description of the immense flocks of wild fowl, chiefly pintail, pochard and gadwall ducks, which arrive in the marshes of Lake Lob-Nor in February, and rest there for a week or two before travelling north to the taiga and tundra. He observed that these migrants all reached the Tarim basin from the west-south-west.

lower jaw is so articulated as to allow plenty of lateral play. The beak is thrust between the scales of a fircone, and as the mouth opens this lateral motion is brought to bear in such a way that the scale is forced aside and the seed lodged at its base is deftly extracted. This modification of the bill does not appear until comparatively late in development, and the mandibles even of the nestlings are nearly straight. Crossbills of different species range right across the northern forest belt of both Old and New Worlds, and one form is indigenous to Scotland. However, it has been long known that, with fair regularity every three to ten years, the crossbills from Northern Europe invade Britain in considerable numbers, and that some of the birds remain to breed in this country. Thus the continental crossbill is extending its range, and now nests annually in Norfolk and Suffolk and other counties where conditions are suitable. The earliest crossbill invasion recorded is that mentioned by Matthew Paris in 1251, when the birds appeared in numbers in the south of England, and ravaged the orchards for the sake of the apple pips. The waxwing is a more irregular visitor, and one which from the Middle Ages onwards has attracted attention by its quaint and handsome plumage. It was formerly regarded as the harbinger of war or pestilence, and it is a curious coincidence that waxwings invaded Britain in considerable numbers in the winter of 1913–14. The breeding grounds of the waxwing in the heart of the conifer forests of northern Europe and Asia were not discovered until 1856, when Wolley, after long search, obtained authenticated eggs from Lapland; but the bird seems to be as erratic in its nesting haunts as in its migrations, and frequently shifts its quarters.

Certain taiga mammals migrate at the beginning of winter. Deer move southwards to the fringe of the forest zone, and wolves and gluttons follow the herds. The hares winter chiefly along the river banks, where they eke out existence by gnawing the bark of deciduous trees (4). But with these exceptions, hibernation is the rule among the vegetable eaters of the taiga.

Hibernation is a complicated physiological process which is not yet thoroughly understood. The general condition of a hibernating animal resembles that of sleep, only intensified.

There are, however, degrees of hibernation. Some forms, such as hedgehogs and dormice, are completely torpid, while others, such as the squirrels, wake at intervals.

The hibernating mammal becomes poikolothermal, that is to say, the power of maintaining the normal temperature is lost and there is a fall to that of the surrounding atmosphere. It is said that in bats the ordinary daily sleep is accompanied by a fall of the temperature to that of the air. Respiration is exceedingly slow, and sometimes ceases altogether. The oxygen consumption is very low. The oxygen intake of a hibernating marmot is but one-thirtieth of that of the active animal, and of this amount only two-fifths appear in the CO_2 discharged. Dubois found that while the oxygen in the arteries was normal, that in the veins was less than normal, and that both arteries and veins contained an excessive amount of CO_2. An active hedgehog expires after three minutes' submersion in water; but a hibernating hedgehog survives after twenty minutes.

Excretion and digestion almost or entirely cease. The brown bear has the curious habit of filling the intestine with a plug of undigested pine-needles (called in Sweden *Tappen*) which is not voided until the spring. Some forms lay up stores of food which are devoured in the waking intervals. Preble [8] found that the American squirrel sometimes hibernates in a pile of fircone scales, which thus afford it both food and shelter; and Loring ([8] cited) also saw quantities of mushrooms —half a bushel in one instance—accumulated near the winter nest. One of the most startling phenomena in hibernation is the sudden rise in temperature that accompanies the awakening. Pembrey and Hale [10] have shown that the temperature of a bat may rise 17° F. in fifteen minutes, and that of a dormouse may rise 22° F. in an hour.

The old view that hibernation is due only to cold is no longer tenable, for it has been shown that when dormice are kept below a certain temperature the animals remain active, but when cottonwool is packed around the cages to give warmth, hibernation may begin. In the Adirondacks the woodchucks become torpid at the autumn equinox when the weather is mild, but they awaken in spring in snow and frost. The marmot also has been known to hibernate in summer-time. Hence we are justified in believing that environmental conditions by themselves cannot induce a winter sleep.

Nor is the condition due only to food, though this plays its part, for hibernating animals are usually very fat at the beginning of winter. It is said that in winter bears do not hibernate unless the summer feeding has been good enough to allow them to lay up sufficient reserves. This is probably one reason why mammals in the tundra remain active in the cold season. Where the summer is so short there is never more than just enough food to go round, and none to put by for a time of scarcity. Thus those forms which must hibernate or perish have not penetrated into the high north.

The most reasonable view seems to be that the power to hibernate, though partly controlled by external conditions, is inherent in some animals and not in others; but so far attempts to locate the stimulus in one organ or another have met with no success.

The capacity for hibernation is distributed rather arbitrarily. Thus susliks, squirrels, marmots, chipmunks, and the jumping rabbit (*Alactaga*) hibernate, but not rats, mice, lemmings, beavers, hares, nor porcupines. Hedgehogs hibernate, but not moles nor shrews.

It is said that formerly in the Russian province of Pskov, the peasants used to winter in a state approaching that of hibernation. Food was scarce, and to conserve bodily energy as far as possible the whole family used to sleep on the warm stove for the greater part of the twenty-four hours, only waking to eat a morsel of bread and replenish the fire. In this semi-dormant condition they passed four or five months.

In the taiga, wind is not the important factor that it is in tundra or steppe, where as the *purga*, or *buran*, it adds greatly to the severity of the winter climate. But indirectly wind is of great œcological significance in forest life. It has been pointed out that plants with juicy fleshy seeds are common in the conifer zone; and it has been suggested that in the calm of the forest dissemination of seeds by wind is impossible, and that dispersal by frugivorous birds and mammals has taken its place. This idea has some support from the number of seed-eating birds on the taiga list. Some doubtless eat insects also, but, on the whole, insectivorous birds, such as wagtails, fly-catchers, hirundines and warblers are relatively scarce, and there is a proportionately large number of seed and berry-eating forms, such as finches, thrushes and buntings. In this respect the northern forests differ from those of New Zealand, where, according to Myers and Atkinson(7), out of thirty-six indigenous forest birds, twenty-eight subsist either wholly or partly upon insects. Woodpeckers, and perhaps cuckoos, are the exceptions to the seed-eating majority in the taiga. The former are cosmopolitan except in Madagascar and Australasia, and their absence is already felt by the foresters in Australia. Froggatt(5) remarks that the forests of Australia appear more liable to insect attack than forests in other parts of the world, owing to the absence of woodpeckers. The only bird in Australia capable of destroying insects embedded inside bark seems to be the black cockatoo (*Calyptorhynchus funereus*).

Wind has an indirect influence on the life of the forest by spreading fires, which often devastate enormous areas. When travelling on the Yenisei, in 1914, my steamer passed for a distance of sixty versts under a thick pall of smoke from one of these fires, which was invisible below the horizon. Such a con-flagration may burn for days, or even weeks, and destroy thousands of acres of trees; for it is nobody's business to fight the flames, even if such efforts would avail. The destruction of the timber changes the face of the country for many years to come. The charred area is gradually overgrown with saplings and low plants and berry bushes, which, according to circum-stances, attract or repel various forms of animal life from the forest around. Then the trees gain supremacy again, and the pendulum of life swings back. The undergrowth is starved out, the burnt ground is buried anew under barren conifer needles, and the fauna is adjusted to the former conditions. As in other parts of the sub-arctic wooded zone, fire accounts for much of the scarcity of virgin forest in the sense in which the term is used elsewhere. Great areas which have not been touched by man are nevertheless of secondary growth, because flames at one time or another have consumed the older trees and their accumulated débris.

Apart from the cleared tracts in the south where agriculture is carried on, the taiga in the region north of the Trans-Siberian Railway has been extensively felled round settlements and along the river banks. The forest behind these artificial clearings is as yet hardly touched by civilised man; but that this will continue is not to be expected, and the trapper and hunter have already made great inroads on the native fauna. Nevertheless, the Siberian taiga at the present day may still claim to be one of the largest and least known wildernesses left in the world.

BIBLIOGRAPHY

(1) ALLEN, M. GLOVER (1923). "Geographic Distribution of certain New England Mammals." *Amer. Nat.* vol. LVII, No. 649.

(2) BEEBE, C. WILLIAM (1917). "Tropical Wild Life." New York.

(3) BREHM, A. E. (1896). "From North Pole to the Equator." London.

(4) CZAPLICKA, M. A. (1916). "My Siberian Year." London.

(5) FROGGATT, W. W. (1923). "Forest Insects of Australia." Sydney.

(6) HUDSON, W. H. (1903). "The Naturalist in La Plata." London.

(7) MYERS, J. G. and ATKINSON, E. (1923). "The Relation of Birds to Agriculture in New Zealand." *N.Z. Journ. Agric.* vol. XXVI, No. 5.

(8) PREBLE, EDWARD A. (1908). "A Biological Investigation of the Athabasca-Mackenzie Region." *N. American Fauna, U.S. Dept. Agric. Biol. Survey*, No. 27.

(9) PREJEVALSKY, N. (1879). "From Kulja across the Tian-Shan to Lob-Nor." Trans. E. Delmar Morgan. London.

(10) SCHÄFER, E. A. (1898). "Text-book of Physiology." London. (Section on "Animal Heat" by M. S. Pembrey.)

(11) SETON, E. THOMPSON. "The Arctic Prairies."

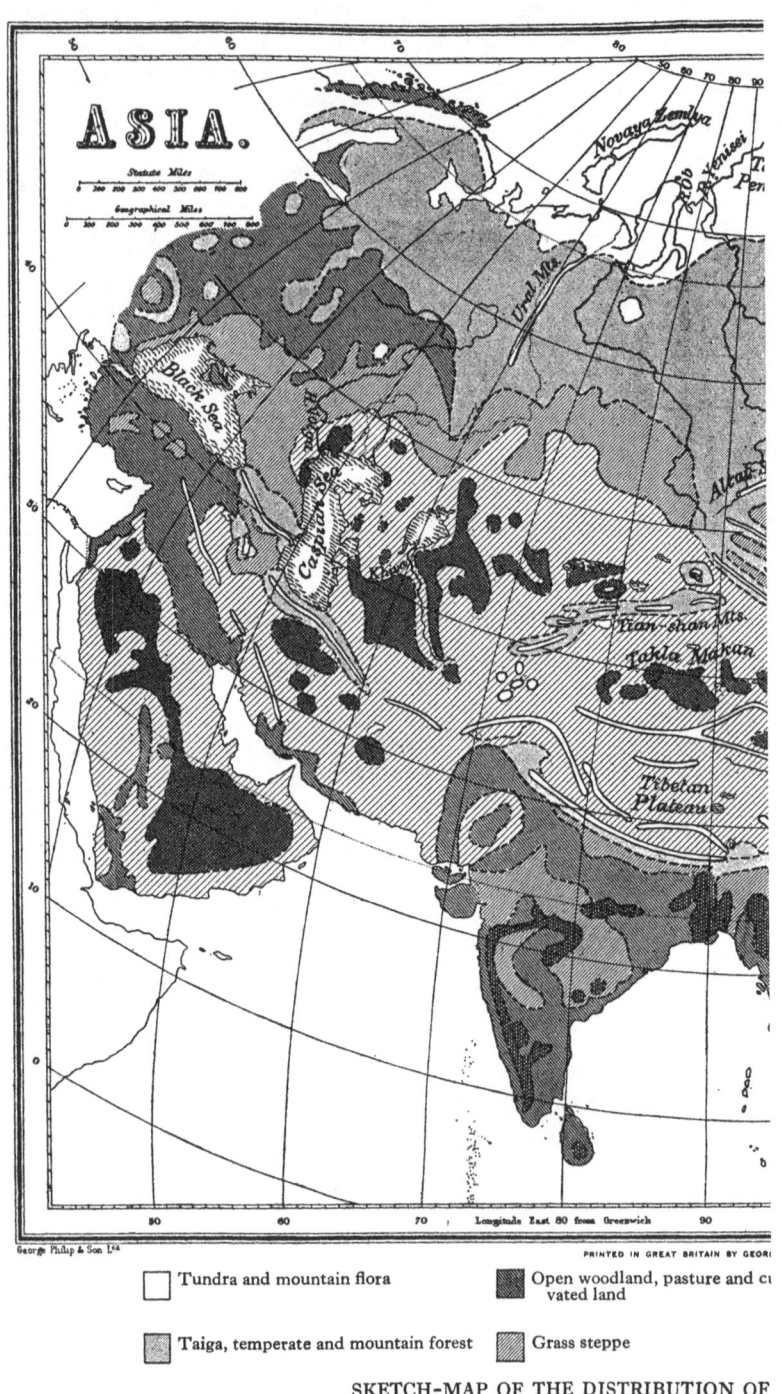

☐ Tundra and mountain flora	■ Open woodland, pasture and cu vated land
▨ Taiga, temperate and mountain forest	▨ Grass steppe

SKETCH-MAP OF THE DISTRIBUTION OF

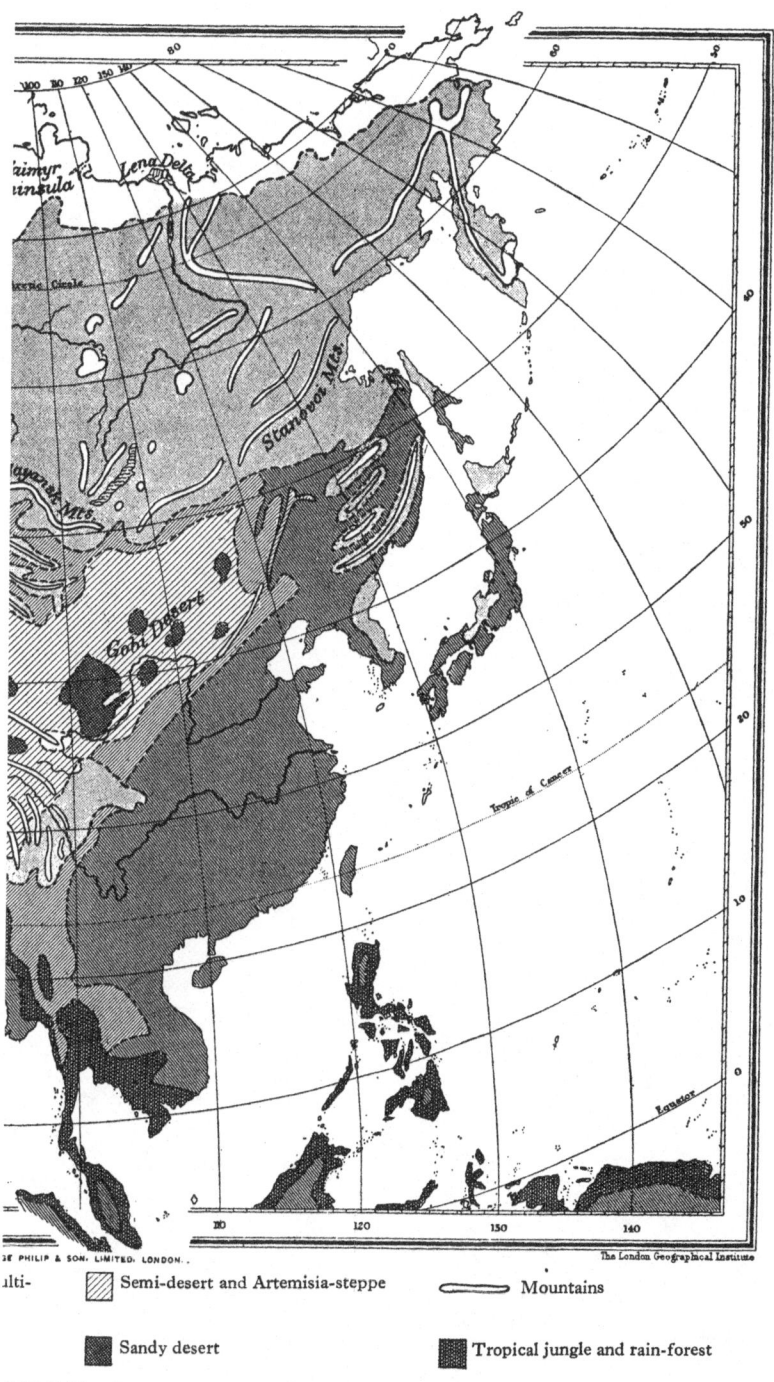

ilti-	▨ Semi-desert and Artemisia-steppe	⬯	Mountains
	■ Sandy desert	■	Tropical jungle and rain-forest

FOREST, STEPPE, AND TUNDRA IN ASIA

INDEX